The winelands of Britain:

Past, present & prospective

Second Edition

Richard C. Selley

Published by Petravin, P. O. Box 425, Dorking, Surrey, RH5 4WA, UK

FRONT COVER: Right foreground: Denbies 100 hectare vineyard and winery, planted in the last century on south-facing chalk slopes of the North Downs. Left background: the Surrey Hills, on whose Greensand scarps vineyards were abandoned in the Little Ice Age.

First Edition 2004
Reprinted 2004, 2005, 2006, 2007 & 2008
Second Edition 2008

Published by Petravin www.petravin.co.uk
P O Box 425
Dorking
Surrey
RH5 4WA

Digitally printed by Vincent Press Ltd. www.vincentpress.co.uk
Vincent Lane
Dorking
Surrey
RH4 3SA

International Standard Book Number: 978-0-9547419-2-1

The winelands of Britain:

Past, present & prospective

Second Edition

by

Richard C. Selley

Dedicated to Adrian White CBE, DL

Owner of Denbies, who turned a geologist's fantasy into a vineyard

"Civilization occurs by geological consent – subject to change without notice."
WILL DURANT (1885-1981)

BOOKS BY THE SAME AUTHOR

Ancient Sedimentary Environments (1970, 4th edition1996)

Introduction to Sedimentology (1976, 2nd edition1982)

Petroleum Geology for Geophysicists & Engineers (1983)

Elements of Petroleum Geology (1985, 2nd edition 1998)

Applied Sedimentology (1988, 2nd edition 2000)

The Box Hill & Mole Valley Book of Geology (2006)

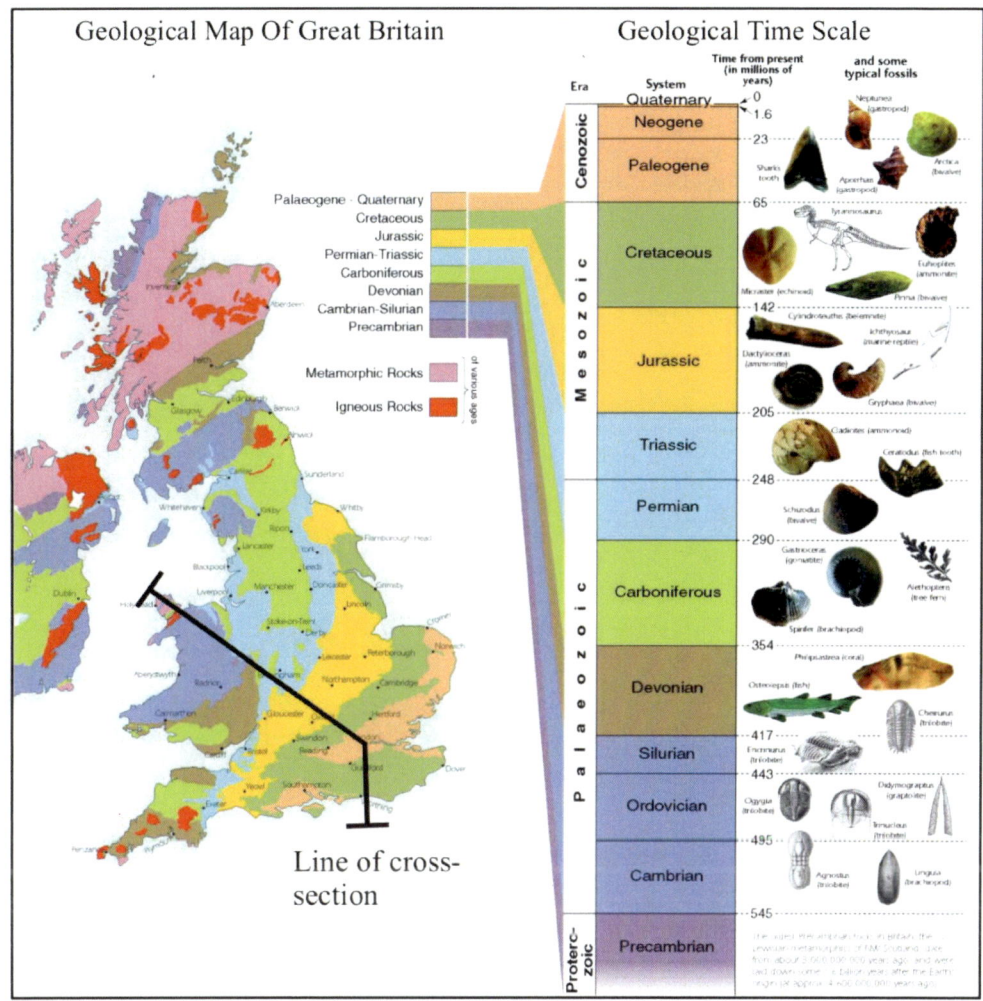

Geological Map Of Great Britain   Geological Time Scale

Line of cross-section

FRONTISPIECE Geological map of Great Britain and Geological Time Scale. Courtesy of the School of Earth Science and Geography, Keele University. The line of cross-section is illustrated in Figure 7.5 and is explained on pages 83-86.

Though vines grow across Britain on rocks of many ages and types, the interplay of geology and climate determines the soil in which a vineyard grows and the landscape on which it stands. Vines first colonised Britain over 50 million years ago, and have only retreated during glacial maxima of the present Quaternary Ice Age. *'The Winelands of Britain'* maps the ebb and flow of vineyards across Britain during the last two millennia, and predicts their future distribution, as global warming continues.

TABLE OF CONTENTS

CHAPTER 1. Introduction.

Drink thy wine with a merry heart.
ECCLESIASTES 9. 7.

PART 1: BRITISH WINELANDS: PAST

CHAPTER 2. Viticulture in Roman Britain

The soil can bear all fruit, except the olive, the vine......
TACITUS 97-8 AD

CHAPTER 3. Viticulture in the Medieval Warm Period

This district, too, exhibits a greater number of vineyards than any other county of England, yielding abundant crops and of a superior quality...'
WILLIAM OF MALMESBURY (1095-1143) writing of the Vale of Gloucester

CHAPTER 4. Viticulture in the Little Ice Age

.....produced most excellent good wines, and a very great quantity of them.
DANIEL DEFOE writing of Deepdene vineyard 1724-6

PART 2: BRITISH WINELANDS: PRESENT

CHAPTER 5. Geological controls on viticulture

Civilization occurs by geological consent – subject to change without notice.
WILL DURANT (1885-1981)

CHAPTER 6. Viticulture in the Industrial Revolution Warm Phase

It is a truth universally acknowledged, that an estate with a southerly prospect, must be in want of a vineyard. JAMES AUSTEN 1813.

PART 3: BRITISH WINELANDS: PROSPECTIVE

CHAPTER 7. The future of British viticulture in a changing climate

I fear tundra in Tonbridge, and glaciers in Guildford more than vineyards in Kent...
Letter in the TIMES 1 November 2001

PREFACE TO THE FIRST EDITION

This book describes how for two millennia British vineyards have been controlled by the impact of changing climate on changeless landscapes.

The winelands of Britain begins with an historical account of British viticulture. Fossil evidence shows that vines have been indigenous to the British Isles for most of the last 50 million years. It is only during the glacial spells of the current Ice Age that they have been absent. During warmer interglacial phases vines have thriven. They have been grown domestically for nearly 5,000 years. Wine was imported into Britain from the European mainland in the later Iron Age. Indigenous viticulture began with the Roman conquest. Thereafter viticulture has been endemic, albeit varying in extent with changing climatic and socio-political factors. Evidence of viticulture is rare in the Dark Ages. Many new vineyards were planted after the Norman Conquest. Thereafter viticulture flourished in England and southern Wales throughout the Medieval Warm Period. It declined in the Little Ice Age of the 15th - 19th centuries to become the hobby of eccentric rural gentry in southeast England. Viticulture has undergone a renaissance in the Industrial Revolution Warm Phase of the last century. English vineyards are now re-established over the same terrain as in the Medieval Warm Period, southeast of a line from the Severn to the Humber.

The winelands of Britain then discusses the strong, though indirect, control of geology on viticulture. The interplay of climate and geology controls landscape, drainage, microclimate, soil nutrients and texture. These factors all affect vine growth and wine character. The book shows how geology controls the landscapes of England's ancient and modern vineyards, illustrated with case histories of vineyards from each climatic phase of the last two millennia.

The core of *The winelands of Britain* is a suite of maps that show the location of some 500 vineyards in the Roman Period, the Medieval Warm Period, the Little Ice Age, and the Industrial Revolution Warm Period. This database is used to construct a graph of the relationship between vineyards and temperature, and to map the ebb and flow of vineyards across the British landscape since Roman times. Integrating vineyard chronology and geological location delineates 4 types of British winelands: abandoned winelands (e.g the Greensand Hills of Surrey), abandoned winelands re-born (e.g. the Chalk Downlands and the Thames Valley terrace gravels), virginal winelands (e.g. the Central Weald), and future winelands - should global warming continue. According to current models of climate change, within decades vineyards may be planted on the southern slopes of the Peak District, and then the Lake District. By the turn of the century vineyards may thrive on the Southern Uplands and Grampian Mountains of Scotland. With the decline of traditional agriculture, viticulture offers a more congenial and environment-friendly use of land than golf courses, and if it fails one can always drink one's liquid assets.

I have written this book hoping that it will entertain the professional scientist, and both entertain and inform all who are interested in the history of the Earth in general, of the landscape of the British Isles in particular.

Richard Selley

1 April 2004

PREFACE TO THE SECOND EDITION

The First Edition of 'The Winelands of Britain' was written in the afterglow of the summer of 2003. The research described in this book began as a study of the role of geology on UK viticulture. Recognition of the increasing importance of climate change diverted research towards the interplay of geology and climate on viticulture. The data in the first edition appeared to support the thesis that the northern limit of viticulture moved to and fro correlative with temperature during the last two millennia. This thesis survives unchallenged.

Since the publication of the First Edition the Intergovernmental Panel on Climate Change has published a further report (2007). The evidence for global warming, and the role of mankind in its creation has received further corroboration. The Second Edition reflects this change of emphasis. The First Edition concluded by speculating whimsically on the winelands of the Côtes D'Ecosse in 2100. This prompted the Daily Express to run the headline 'Drink Chateau Loch Ness for a monster hangover'. Predictions of the rate and regional variation of temperature change have now become so refined, however, that it is possible to delineate both the location of the future winelands of Britain, and also the varieties of grape that they may grow in the coming decades.

The need for a new edition to update the predictions of the impact of climate change on British viticulture provided the opportunity to update references, embellish the text and illustrations, and to correct those trivial errors that thoughtful readers drew to my attention.

1 April 2008

ACKNOWLEDGEMENTS

The author gratefully acknowledges the many geologists and vineyard owners in England and abroad, South Africa and Australia in particular, who have shared their knowledge, insights and wine during the preparation of this work.

Specific thanks are due to those who critically reviewed the manuscript, pointing out errors, and suggesting improvements. These include Mr Stephen Skelton, Mr Adrian White and the late Professor Jake Hancock, who was particularly helpful in providing the historical details of the Cognac 'bulls eye' *'jeste géologique'*.

The sad saga of the IX Hispana Legion, which was not only lost, but has also lost its vineyard at Lindum (Lincoln), was unravelled with the help of the Very Reverend Alec Knight, Dean of Lincoln Cathedral, Mr Andrew Taylor, Chief Executive of Lincoln Town Council, Ms Kate Fenn, Civic and Twinning Officer, and Dr Glynn Coppack, Inspector of Ancient Monuments for English Heritage, who finally laid the vineyard to rest.

Dr P Kenrick of the Natural History Museum helped me with the palaeobotany of vines, and several of his NHM colleagues tracked down the quote from Will Durant, whose provenance had escaped me. Dr Jonathan Williams of the British Museum introduced me to the vinous coinage of the Celtic kings.

Malcolm Lewis, formerly of Media Projects International, kindly allowed me to use the photograph on the back cover, and Ian Meadows the photograph of the Roman vineyard at Wollaston.

I am grateful to Chris Foss, and his colleagues and students at Plumpton College, who have enabled me to test drive and develop some of the ideas in this book

Thirty-five years ago my first book took twelve months from delivery of typescript to publication. Technology has advanced dramatically since then. Eight years ago however, one of my books took the same time, even though the text was delivered on disc and most of the illustrations were already drafted. I have suffered from inept promotional copy drafted, apparently, by school children on work experience, and from 'printer's devils' not of my own making. I have had books stuck 'in press' for years while publishers were taken over and re-organised. Thus it appealed to me to publish this work myself, particularly since it is written in a scholarly style, though with personal whimsies that would never survive peer review by my academic colleagues. I typed the text on my PC. Dr Alex Davis, a former student, crafted the illustrations. My son-in-law John Gibson tweaked the graphs. My daughter Andrea designed the jacket and taught me desktop publishing. My wife forgave me for breaking my promise to never, never, never ever write another book. Then the work was delivered on a CD to my friendly local printer who ran it off in a fortnight. What could be simpler?

My experience since publishing the first edition has confirmed my views. With digital 'print on demand' all I have to do when stocks run low is to email my printer to run off some more copies. They arrive at my door within days.

NOTES ON UNITS

As far as possible SI units have been used throughout the book, accompanied by British and other units where appropriate. Thus:

Area is in *Hectares* (1 hectare = 2.47 Acres)

The *Arpent* is a Norman-French measure of area widely used to record the size of vineyards in the Domesday Book (see p.22). It is between 1.25 and 0.8 acres. No longer used in Europe, the Arpent was exported to the New World, and was still in use in some former French colonial possessions (Quebec, Louisiana) in the last century.

Length is in *kilometres* (1 kilometre = 0.62 of a mile)

Volume is in *Litres* (1 litre = 1.76 pints)

Modern standard wine bottles contain 0.75l, or 75cc. (27 Fluid ounces). The *Muid* is an improbable-sounding liquid measure used in Saxon times (see p. 22). It is believed to approximate to 164 litres (36 gallons). The *wineskin* has been a measure of volume since Biblical times. The skins of various animals were deemed fit for this purpose, but apparently a goat's scrotum was the container of choice (see p.21). Unfortunately the local Trading Standards Office was less than helpful when asked for the metric equivalent.

Yield of a vineyard can be measured in terms of the weight of grapes for a given area (tons/acre, or tons/hectare, etc. in the New World). More usually it is measured in terms of the volume of wine produced per unit area, commonly *hectolitres per hectare*, expressed as hl/ha. (A hectolitre is 10^2 litres) A useful indicator of the productivity of medieval vineyards is the Bm/pa (number of Benedictine monks per annum that a vineyard can sustain). Given that under the Rules of St Benedict a monk is only allowed to enjoy 1 pint (0.56l) of wine per day, it follows that a monk can enjoy 365 pints (204.4l) a year. The yield of a vineyard can thus be expressed in Bm/pa (see p.22).

CHAPTER 1.

Introduction.

Drink thy wine with a merry heart.
ECCLESIASTES 9. 7.

1.1 IN THE BEGINNING

Yeast is a primitive microscopic plant that plays an important part in civilization in general, and conviviality in particular. The variety used in making bread, beer and wine is termed *Saccharomyces cerevisiae*. This occurs wild in nature. Yeast has the ability to break down sugar to form alcohol (ethyl, not the methyl alcohol that causes blindness) and carbon dioxide gas – the fizz in beer and sparkling wines. The chemical reaction may be expressed as

$$C_6H_{12}O_6 = 2C_2H_5OH + 2CO_2$$
Sugar → alcohol + carbon dioxide

As it drifts around in the wild yeast thrives on overripe fruit, turning its sugar into alcohol and releasing carbon dioxide gas. This process gives the zing on the tip of the tongue produced by biting into an over-the-top piece of fruit. Not only may fermented fruit juice produce a pleasure on the palate, but also, when imbibed and digested, it produces a pleasurable sensation. Over indulgence however, produces many very unpleasant symptoms, including loss of bodily and mental functions.

The pleasures of imbibing alcohol are known throughout the animal kingdom, from bees to elephants. Stories are legion, of boozy bees too confused to find their way back to their hive, monkeys incoherent from overindulging in fermented figs, convivial elephants leaning against trees like bookends, and elkoholic elks (Fig.1.1).

In the whole animal kingdom only starlings appear to be immune to alcohol, having an enzyme so powerful that it can break down alcohol 14 times faster than humans. This explains why no one has seen an intoxicated starling. Since modern monkeys and apes enjoy alcohol, it is reasonable to suppose that our primate ancestors did too, and that the history of alcohol-induced hominid happiness stretches way back into prehistory.

Figure 1.1 Enjoyment of fermented fruit juice is ubiquitous in the animal kingdom.

Illustration by Guy Troughton, from BATS by Phil Richardson (©Whittet Books, Stowmarket: www.whittetbooks.com)

1.2 THE ORIGIN OF WINE

In its broadest sense the word 'wine' is given to the fermented juice of any fruit. Conventionally, however, it is applied to the fermented juice of the grape, unless prefixed, as in barley wine, or given specific names, such as, cider or perry, for fermented apple and pear juice respectively. In this book 'wine' is used in its restricted sense for the fermented juice of the grape vine, *Vitis vitifera*. Traditionally a vital distinction has been made between the terms British wine and English wine. British wine has been produced by the fermentation of grape juice imported to the UK from who knows where. English wine, however, is made from English grapes. Welsh wine is made from Welsh grapes, and, if global warming continues, Scottish wine will be made from Scottish grapes. Because wine is currently made in England, Wales, Northern Ireland and the Channel Islands (and may soon be made in Scotland), the term British wine may take on a new and more honourable meaning,

Vines (the genus *Vitis*) evolved way back in the geological past. Fossilised vine pollen, pips, leaves and twigs have been found in mud deposited in the London and Hampshire basins some 55 million years ago (Reid and Chandler, 1933). It is probably true to say that vines have been indigenous to Britain thereafter. Only during the Ice Age of the last 2 million years, have vines retreated to the warmer south, but only during the glacial maxima, returning episodically during the warmer interglacials (Godwin, 1975). The modern grape vine, *Vitis vitifera*, is believed to have originated in what is now Georgia and Armenia, between the Black and Caspian seas (Fig. 1.2). Largely by human means it migrated westwards into the Mediterranean region during late prehistoric times (Nunez and Walker, 1989). Though most vines enjoy a tropical climate, the grape vine prefers a Mediterranean climate with temperatures of between 10 - 20°C. There is some evidence that winemaking began in the Middle East in the Neolithic phase of human culture some 7.5 thousand years ago. Certainly wine presses existed in Egypt some 5 thousand years ago. Viticulture was well established throughout the Middle East and the Mediterranean littoral by Classical times. There is an extensive ancient literature on wine, its processing and consumption.

For millennia wine has played an important part in society in general and in medicine and religion in particular. The Greek doctor Hippocrates of Chios (460-377 BC), and his later Roman counterpart, Galen (129-216 AD), both recommended wine for internal use as a medicine, and externally as a disinfectant for open wounds.

The bible contains many references to vineyards and winemaking. Noah planted a vineyard as soon as the flood receded (Genesis 10.20). The resultant family fracas was just the sort of behaviour that led Mohammed to forbid Muslims to drink alcohol. The Good Samaritan used wine to disinfect the wounds of the mugged man (Luke 10. 34). Jesus' first miracle was to turn water into wine (John, 2), a method of celestial vinification that the church has failed to replicate in two millennia. St Paul urged Christians *'Drink no longer water, use a little wine for thy stomach's sake, and for thine often infirmities'*. (1. Timothy 5. 23) The medical application of the vine is taken to the extreme at Les Sources de Caudalie in Bordeaux (www.sources-caudalie.com). This health spa offers *'vinotherapie'*. Grape skins and pips are used in various body massages and scrubs.

Fig. 1.2 *Vitis vinifera, the common vine.*

1. *Fruit*

2. *Young flower*

3. *Cross-section of flower.*

Wine has traditionally played an integral part in many religions. To the ancient Greeks and Romans wine was considered a gift of the gods. Wine symbolised the blood of the Earth, and was deemed a suitable alternative sacrifice to the blood of animals. A libation of wine was poured out onto the ground to placate the gods on ceremonial occasions, a custom that continues to modern ship launching ceremonies. The Iliad and the Aeniad provide detailed descriptions of the use of wine in funeral rites, both for extinguishing funeral pyres and for bone-washing. The Greek god Dionysus, and his Roman counterpart Bacchus, was a patron saint of wine and conviviality, often excessive. The festivities of the Bacchanalia were an early form of hen party, a good night out for the girls.

For Jews wine plays an important part in ceremonies to mark all of life's stages, from circumcision to burial. It is particularly important during the religious festival of the Passover. This includes the Seder, a ceremonial meal, in which 4 glasses of wine are drunk at carefully prescribed intervals, and sometimes a fifth poured and left untouched to await the arrival of the Messiah. The Christian ceremony of Holy Communion is derived from the Passover meal of the last supper. Its significance is missed without knowledge of the Seder from which it sprang. Today people of many faiths and none say 'Cheers', or words to that effect, before drinking; an atavistic hangover of past religious beliefs.

Finally, who can disagree with the quaint encomium of Nicholas Culpepper (1616 - 54), seminal herbalist and pharmacist, who wrote of the vine: *'it is a most gallant tree of the Sun, very sympathetical to the body of man; and that is the reason spirit of wine is the greatest cordial among all vegetables'* (1653 – 1809).

1.3 WHAT THIS BOOK IS, AND IS NOT, ABOUT

Wine, in all its aspects from viticulture to the role of wine in society, medicine and religion, has generated a huge literature over the millennia. British, more specifically English, vineyards, ancient and modern, have been described in detail, though not as much as those of France and some New World countries. Hyams (1949), Ordish (1953, 1977) and Barty-King (1977, 1989) have delineated the history of British viticulture over the last two millennia in authoritative detail. Skelton (2001) has written an excellent lexicon of modern British and Irish vineyards that includes a concise account of the history of British viticulture.

This book is also about viticulture in Britain, including England, Wales and Scotland. It approaches the subject, however, in the context of the impact of climate change and of geology on viticulture. The climate of the earth has been constantly changing. It has undergone a gradual cooling over the last 60 million years, and abrupt cooling during the Pleistocene Ice Age of the last two million years. During the Ice Age there have been warm spells, however, and we are enjoying one now. The last 10,000 years have been a period of quite unusual climatic stability. Nonetheless fluctuations have been noted. The Roman and Medieval periods were relatively warm, with an intervening thermal sag in the Dark Ages, and the cooler 'Little Ice Age' of the 15th - 19th centuries. With the advent of the Industrial Revolution, the climate has warmed again to reach temperatures comparable to those of the Roman period.

Saserna (in Columella, 67AD) theorised that viticulture was a marker for climate change some two millennia ago. This book examines the relationship between viticulture and climate, past, present, and prospective. To do this it necessarily relies on, and reproduces summaries of, previously published work on British vineyards ancient and modern.

There is an old geological cliché 'the present is the key to the past'. It is often metamorphosed into 'the past is the key to the future'. Using this precept this book then considers the impact of climate change on the future of British viticulture. If the predictions of global warming are correct, then the future of British viticulture is bright.

Whereas now it flourishes in southern England, and copes with the climate of the Midlands, in future decades viticulture may gradually migrate ever northwards, from the southern slopes of the Derbyshire Peak District to the Southern Uplands of Scotland, and then to those of the Grampians. Not only may the future of viticulture in Britain be bright for wine making, itself, but wine tourism may also blossom. This is an important variety of tourism in both the Old World and the New (Hall et al. 2000). Wine tourism is already developing in England (Howley and Westering, 2000). There are several wine trails for tourists to follow in the Weald and the West Country. Since as far back as the last century various learned geological societies have run field trips to study the geology of English vineyards. The thirst for these trails and trips is growing, and wineries should endeavour to quench it.

The second theme of this book is to examine how vineyards are established in the British landscape. We inhabit this planet by courtesy of its geology. Wealth and civilization are created by farming the earth, or extracting resources from it. Some vineyards are located by serendipity, but most good ones are located with an eye to landscape and soil. These result from the inter-reaction of climate, past and present, with geology. This book describes the control of geology on the British landscape and its contained vineyards.

In some instances, modern vineyards have been replanted over ancient ones because of the way in which geology has formed favourable landscape. Whereas the geology of vineyards in general, and French ones in particular, has been described in detail, this is the first book to describe British vineyards in their geological setting. It is not surprising that the effect of geology on whisky and beer is better known. The geology of whisky has been described by Cribb & Cribb, (1998) and Cribb, (2004), and of beer by Cribb (2004a).

1.4 HEALTH AND SAFETY: THE GOOD NEWS AND THE BAD

Current medical opinion holds that alcohol in moderation has health-giving properties, lowering the risk of cardio-vascular disease in particular. Over indulgence, on the other hand has quite the opposite effect. Prolonged over indulgence increases the risk of gastrointestinal cancers, especially when combined with smoking, and of liver disease. Binge drinking is a major cause of accidental injuries, in the home and in motor vehicles, and of many violent crimes. Excessive intake of alcohol over a short time span can cause death from alcoholic poisoning. Current medical opinion seems to be that the correct dose to gain the optimum benefit of alcohol is about 3 - 4 units per day for men and 2 - 3 for ladies. A unit is defined as:

$$1 \text{ unit} = \frac{V \times A}{1000}$$

Where V = the volume of the liquid (ml).
A = the % alcohol by volume

A small (125ml) glass of red wine with an alcohol content of 13.5% calculates out at 1.69 units. A 125ml glass of white wine, with a lower content of 12.5%, is 1.56 units. English wines have alcohol contents of around 11 and 10% for red and white wines, calculating out at 1.38 and 1.25 units respectively. Thus 3 glasses of English wine, equates with 3.37 units, the optimum for good health. Applying statistical gymnastics and assuming a Gaussian distribution, it seems that 4 glasses of wine a day is equivalent to drinking 2 glasses, 5 glasses the same as 1. Drinking 6 glasses of wine a day has the same health benefit as teetotality (Figure 1.3). Cheers.

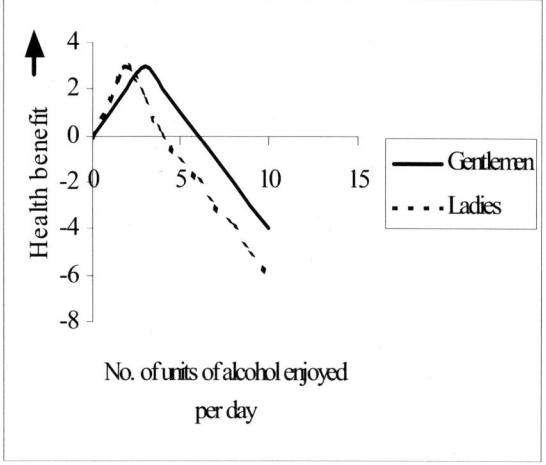

Figure 1.3 Graph to show the health benefit (vertical axis) for every unit of alcohol enjoyed per day (horizontal axis). Since, for a gentleman enjoying 3 units a day obtains the optimum health benefit, it follows that 6 units a day is as healthy as teetotality. For a lady the optimum health benefit is obtained by enjoying 2 units a day. It therefore follows that 4 units a day is as healthy as teetotality.

The value of wine as a medicine has been cherished since the beginning of time. There is no doubt that this field of research reached its apotheosis in the work of Dr E A Maury. Dr Maury graduated from the Faculté de Médecine de Paris. He was a general medical practitioner, an acupuncturist, and homeopath. At one time Dr Maury was a Resident Specialist at the Royal Homeopathic Hospital in London. Dr Maury developed a detailed list of the type of wine and correct dosage for almost every disease known to mankind (Maury, 1974). An abbreviated summary of his recommended medicines is shown in Table 1.1. Though Dr Maury explains the rationale behind his choice of wines for different diseases, there does not seem to be any experimental test results to validate his recommendations. Most of his dosages recommend two glasses of wine with a meal. If one only ate a single meal a day this might do no harm. But two glasses of wine with breakfast, lunch and dinner would exceed today's currently accepted healthy limit. He recommends no wine for cirrhosis of the liver.

This author does not accept any legal liability for any adverse effects experienced by readers who follows Dr Maury's recommended medicines.

Social commentators have noted that there is a major trans-European cultural divide. In the north the culture of the Scandinavian, Saxon and Celtic tribes is based on butter and beer. Binge drinking is the usual method of (over) indulging in alcohol. Consider the beery singing rugby players in their clubhouse on a Saturday night. They are direct cultural descendants of Beowulf and his mead-muddled Saxon warriors who caroused the night away in their hall after slaying the monster Grendal. By contrast the life style of the Mediterranean littoral is based on olive oil and wine. Wine is enjoyed in moderation as a gentle activity associated with good food and family. This cultural divide was noted two millennia ago by Roman writers (e.g. Tacitus, 97-98) who commented on how the Celts refreshed themselves with wine and other alcoholic beverages for days and nights on end. A European Union committee has yet to address the problem of this north-south cultural chasm.

MEDICAL CONDITION	RECOMMENDED WINE	RECOMMENDED DOSE
Abdominal distension	Alsace	1 or 2 glasses after meals
Aerophagia	Dry Champagne	2 glasses after meals
Allergy	Medoc	2 glasses after meals
Anaemia	Graves	2 glasses after meals
Anxiety	Médoc	2 glasses with meals
Arteriosclerosis	Muscadet	2 glasses with meals
Arthritis	Saumur	2 glasses after meals
Bronchi – disease of	Bordeaux or Burgundy	3 coffee cups full per day
Cholesterol	Muscadet	2 glasses after meals
Coliform infections	Dry White, or Champagne	2 glasses per meal
Colitis	Low alcohol white	2 glasses per meal
Constipation	Anjou or Vouvray	2 glasses per meal
Convalescence	Médoc or Roussillon	1 glass before & after meals
Coronary diseases	Dry Champagne	2 glasses before meals
Cystitis	Sweet Anjou	1 or 2 glasses per meal
Demineralisation	Châteauneuf-du-pape	1 or 2 glasses after meals
Diabetes	Young red wine	800 grams per day
Diarrhoea - acute	Young Beaujolais	1 glass before & after meals
Diarrhoea - chronic	Old red, such as Médoc	1 glass before & after meals
Dyspepsia	Anjou or Vouvray	1 or 2 glasses after meals
Eczema	Light red or white wine	500cc per day
Enteritis	Médoc, old & light	2 glasses per meal
Eyes (diseases of)	Red Burgundy or Bordeaux	2 glasses per meal
Fatigue	St Emilion/Côte de Beaune	1 glass before & after meals
Fever	Dry Champagne	1 bottle per day
Flatulence	White Alsace	Half a bottle per day
Gout	Sancerre	2 glasses per meal
Haemorrhage	Red Burgundy or Bordeaux	Half a bottle per day
Hypertension	Sancerre	2 glasses per meal
Infarct (tendency towards)	Champagne	1 or 2 glasses with meals
Influenza	Côtes de Rhône	Half a bottle per day
Liver (sluggishness of)	Dry Champagne	2 glasses with/after a meal
Loss of weight	Côte de Beaune	2 glasses per meal
Menopause	Médoc	2 glasses per meal
Nephritis	Light red Bordeaux	2 glasses per meal
Nervous depression	Médoc	1 or 2 glasses during a meal
Neurasthenia	Blanquette de Limoux	Half a bottle a day
Obesity	Provence Rosé or Sancerre	1 bottle per day
Old Age	Burgundy or Champagne	2 glasses per meal
Osteoporosis	Médoc or Côte de Nuits	2 glasses per meal
Pregnancy	Light red Bordeaux	2 glasses per meal
Salmonella & typhoid	Old Red Médoc	Half a bottle per day
Slimming diet	Côtes d'Or	2 glasses per meal

Table 1.1 *Wine and dosage for medical conditions recommended by Dr Maury (1974)*

7

1.5 REFERENCES

Barty-King, H. 1977. A Tradition of English Wine. Oxford Illustrated Press. Oxford. 250pp.

Barty-King, H. 1989. A Taste of English Wine. Pelham Books. London. 209pp.

Cribb, S. & Cribb, J. 1998. Whisky on the Rocks. British Geological Survey. Keyworth. 72pp.

Cribb, S. 2004. The Geology of Whisky. & The Geology of Beer In: Encyclopedia of Geology. (Selley, R.C., Cocks, L. R. M. and I R. Plimer eds.). Elsevier. Amsterdam.

Culpepper, N. 1653. (Editions published up until 1809) The English Physician Enlarged.

Godwin, H. 1975. The history of the British flora: a factual basis for phytogeography. Cambridge University Press. Cambridge. 541pp.

Hall, C.M., Sharples, L, Camborne, B. and N. Macionis (Eds.) 2000. Wine Tourism around the World. Butterworth Heinemann. London. 348pp.

Howley, M. and Van Westering, J. 2000. Wine Tourism in the United Kingdom. In: Wine Tourism around the World. Butterworth Heinemann. London. 175-189.

Hyams, E. 1949. The Grape Vine in England. John Lane. London. 209pp.

Jackson, R.S. 2000. Wine Science: Principles, Practice, Perception (2nd.Edn.). Academic Press. San Diego. 648pp.

Maury, E.A. 1974. Wine is the Best Medicine. Corgi. London 127pp.

Nunez, R.D. and Walker, M.J.C. 1989. A review of the palaeontological findings of early *Vitis* in the Mediterranean. Review of Palaeobotany and Palynology, 61 (3-4). 205-237.

Ordish, G, 1953. Wine Growing in England. Rupert Hart-Davis. London.

Ordish, G. 1977. Vineyards in England and Wales. Faber & Faber. London.

Price, P. V. 2002. Curiosities of Wine. Sutton Publishing. Stroud. 196pp.

Reid, E.M and Chandler, M.E.J. 1933. The flora of the London Clay. (Natural History) London.

Robinson, J, (Ed.) 1999. The Oxford Companion to Wine. (2nd Edn.) Oxford University Press. Oxford. 820pp.

Skelton, S. 2001. The Wines of Britain & Ireland. Faber & Faber. London. 531pp.

Tacitus, P.C. 97-98. Germania. (Published by Penguin Books. 1948. 175pp.)

Unwin, T. 1991. Wine and the Vine. An Historical Geography of Viticulture and the Wine Trade. Routledge. London. 409pp

PART 1

BRITISH WINELANDS: PAST

CHAPTER 2.

Viticulture in

Roman Britain

The soil can bear all fruit, except the olive, the vine......
TACITUS 97-8 AD

2.1 BEFORE THE ROMANS

Vine pips and pollen have been found in Suffolk sediments deposited some 250,000 years ago during the Mindel-Riss warm interglacial phase (Planchais, 1972-3). Pips of a semi-domesticated grape have been excavated from a Neolithic site dated to 2,700BC (Jones & Legge, 1987). There is, however, at present no written or archaeological evidence for pre-Roman viticulture in Britain. Tacitus in his 'Agricola', written in 97/8 AD, describes Britain at the time of the conquest. He comments on the terrible weather: *'The climate is objectionable, with its frequent rains and mists...'*. As noted above, he writes that *'The soil can bear all produce, except the olive, the vine, and other natives of warmer climes....'*

Though there is, at present, no evidence for indigenous viticulture in pre-Roman Britain, it was practiced by the Celtic Allobroges, who dwelt in the mountains on the southern shores of Lake Geneva before the Roman Conquest.

Despite the absence of indications of pre-Roman viticulture in Britain, there is ample evidence that the Celts enjoyed wine long before the arrival of Rome. Large numbers of empty pre-conquest amphorae have been found left lying around the countryside, especially in Essex, and around Chichester. Some of these amphorae originated in Gaul, others from further away in southern Italy. It has been argued that this does not indubitably prove the import of wine, since amphorae were also used to transport and store olive oil and fish sauce (the tomato ketchup of Roman cuisine). Large quantities of amphorae have been found, however, in high status Iron Age graves. It has been remarked that it seems improbable that a Celtic chieftain would be dispatched into eternity with a year's supply of fish sauce. Wine would surely have been far more congenial.

11

A second line of evidence points to the importance of wine drinking, among the aspirational classes at least, if not amongst the general populace. Coins of late first century BC – early first century AD bear extensive images related to wine drinking (Williams, 1998). These images include the *cantharus*, a Roman wine cup, the *thyrsus*, a curious wand associated with Bacchic rites, bunches of grapes, single vine leaves and sprays. Such logos are found on the coins of Cunobelin and Tasciovanus, kings of the Trinovantes (early Essex man) and Verica, King of the Atrebates, who ruled large areas of the present-day Home Counties (Figure 2.1). Viticultural images are commonplace in Celtic art, causing one expert to comment that most Celtic art is related to drinking. It is improbable that itinerant artists from the Classical world imported these wine-related motifs, since they are absent from the Celtic coins of the intervening lands of Gaul and Germania. It is more probable that they arose spontaneously to indicate the trading power and conviviality of their minting kings.

Figure 2.1 Gold stater with vine leaf logo (right). Minted by Verica, King of the Atrebates fl. 43AD (courtesy of P.Clayton).

It is unlikely that this coin provides evidence of pre-Roman viticulture, more probably it is an indication that Celtic chieftains enjoyed imported wine.

A third line of evidence points to the Celts enjoyment of wine. There are the accounts of contemporary historians, such as Diodorus Siculus, who wrote in the first century BC:

'The Celts partake of this drink without moderation, by reason of their craving for it they fall into a stupor or a state of madness....'

Sadly there is nothing new in British binge drinking, it goes back at least 2000 years.

2.2 THE ROMAN CONQUEST

Subsequent to the Roman conquest of Britain two lines of evidence point to viticulture: literature and archaeology. In or around 90 AD the Emperor Domitian issued an edict ordering the destruction of all vineyards north of the Alps. This edict would have included the province of Britannia, but does not of itself prove that there were vineyards that far north. The Emperor Probus, however, was more specific, lifting the ban on viticulture in Britain and Gaul in 280 AD.

Archaeological evidence shows that vineyards did indeed flourish in Roman Britain, regardless of Imperial edicts. The identification of ancient vineyards in general and of Roman ones in particular, is fraught with problems.

An abandoned vineyard is not a prominent archaeological feature, like a castle or a wall. It may be unrecognisable as such, to layman and trained archaeologist alike, unless they know what to look for. Furthermore, finds of terracing, grape pips, and of pits and post holes suggestive of viticulture, may do just that. But the vineyard in question may have been producing dessert grapes, not necessarily grapes for wine making. Terraces are a common feature of steep slopes from the Downs of southern England to the Yorkshire Dales. Known to geomorphologists as '*terracettes*', they are attributable to the down slope creep of weathered detritus. Hikers may believe that terracettes were produced by contour-hugging sheep, until they note them extending to the feet of dry stone walls and continuing uninterrupted on the other side. Had sheep the athletic ability of kangaroos they might produce terracettes, but as they do not, they cannot. Terracing on steep slopes adjacent to Roman villa sites, as at North Leigh, Oxford, and on a south-facing slope of the North Downs at Sutton Valance, may be abandoned vine terraces, similar to those of Mediterranean hillsides. More convincing is the recent discovery of terracing with *Vitis* pollen found at Dover's Hill near Chipping Camden (O. Pritchard. *Pers. Com.* 2008)

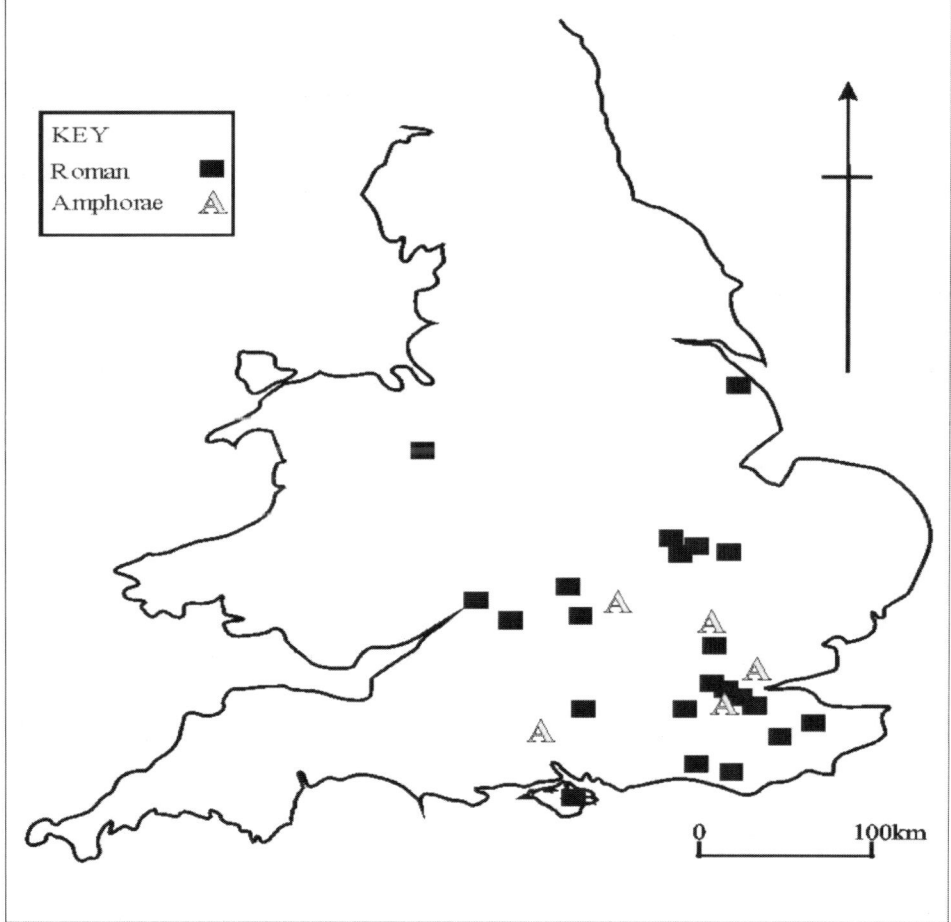

Figure 2.2 *Map of Roman vineyards, including hypertentative sites based on a single grape pip, as well as extensively excavated vineyards. Letter **A** indicates wine amphora factories.*

COUNTY	LOCATION	SOURCE	REMARKS
Bucks.	Stanton Low	Woodfield (1989)	Excavation
Cambs.	Fen Drayton	In Brown *et al.* (2001)	Excavation
Gloucs.	Dover's Hill	O.Pritchard (*Pers. Com.* 2008)	Terraces & *Vitis* pollen
	Gloucester	Godwin (1956), Wilson (1977)	Grape pips, skins & evidence of pressing
Hants.	Silchester	Godwin (1956)	Grape pips
	Adgestone, Isle of Wight	Elusive	South slope adjacent to a Roman villa
Herts.	Boxmoor	Cited in Frere (1978)	Vine plants
Kent	Ightham	Barty-King (1977)	Oral tradition
	Sutton Valance	Sackville-West (1949)	Terracettes
	Wingham	Archaeologica Cantiana	Roman tiles in a field called the vineyard
Lincs.	North Thoresby	Webster *et al.* (1967)	Extensive excavation
	Lincoln	K.Fenn *pers. com.*	Muddled with N. Thoresby G. Coppack *pers. com.*
London	Bermondsey	Godwin (1956)	Grape pips
	City	Godwin (1956)	Grape pips
	Southwark	Godwin (1956)	Grape pips
Northants.	Grendon	Jackson (1991)	Excavation
	Wilby Way	Cotswold Arch. Trust	In Brown *et al.* 2001
	Wollaston	Brown *et al.* (2001)	Extensive excavation
Oxford	North Leigh	Lovegrove in Barty-King (1977)	Terracettes near a Roman villa
Shropshire	Wroxeter	Evidence elusive	Adjacent to Roman city
Surrey	Bagden	Fortescue (1993) & Carr *op cit.*	Oral tradition
Sussex	Cissbury Ring	Turner (1850) & Barty-King (1977)	Excavations & (later) air photos
	East Dean	Barty-King (1977)	Oral tradition

Table 2.1 *Table listing sites identified as Roman vineyards. Note that these include hypertentative sites based on a single grape pip, or less, as well as extensively excavated vineyards.*

Presumed Roman vineyards have been found across Britain to the southeast of a line from the Severn to the Humber (Figure 2.2 and Table 2.1). The most northerly vineyard identified to date is at North Thoresby in Lincolnshire. Here the existence of a vineyard has been inferred from excavations of a system of rectangular ditches associated with Romano-British pottery. The vineyard extended to some 12 acres. Using the figure of 19hl/ha calculated for wine production in Medieval times (see p. 22), the North Thoresby vineyard would have produced some 7,680 litres of wine, say some 10,000 modern 75cc bottles a year. North Thoresby vineyard was located on the eastern side of the Lincolnshire Wolds near Ermine Street. This Roman road ran from Horncastle in the south to Caistor-on-the-Wold in the north. There were several villas strung out along its length.

The major Roman city of Lindum (Lincoln), some 40km to the southwest of North Thoresby, is also reputed to have had a Roman vineyard in the present grounds of the old Bishops' Palace. Had this been true it might have provided refreshment for the IXth Hispana, the mysterious 'lost legion'. Some 30 years ago a vineyard was planted on this site. A quest for the evidence for the Roman vineyard was assisted by the Dean of Lincoln Cathedral, and helpful staff of Lincoln City Council and English Heritage. This revealed that in the 14th century the site was an orchard. This was terraced over in the 1880'ies. Sadly there is no evidence for a Roman vineyard in the grounds of the old Bishop's Palace. Somehow the genuine Roman vineyard at North Thoresby had 'migrated' some 40km SW to accrete onto Roman Lindum. This is an example of how oral traditions develop.

Further evidence for Roman viticulture is provided by home made wine amphorae. These are not identified because they are stamped *'Factare in Britannia',* but because the kilns and/or piles of waste sherds have been discovered at sites in Oxford, the New Forest, St Albans (Verulamium), Brockley Hill and London (Symonds, 2003) Some amphorae have been found coated with pine resin, suggesting that the locals were trying their hand at retsina. A very fine amphora from Moorgate, London, bears the potter's name 'Senecionis' (Seeley & Drummond-Murray, 2005) (Figure 2.3).

Figure 2.3.Wine amphora stamped on the neck with the name of the potter (Senecionis). Made in Britain Dated 1st Century AD From Moorgate, London. © The Museum of London

There is archaeological evidence that viticulture may have been more widespread in Roman Britain than the limited number and distribution of known vineyards may suggest. The *falx vinitoria*, is a curious small iron sickle that was used for pruning vines. This tool was widespread during the days of the Roman Empire and for centuries thereafter. Representations of Bacchus, the Roman god of wine, commonly show him with a bunch of grapes in one hand, and a wine goblet in the other, while skipping or tripping over a *falx vinitoria* at his feet. This scene is depicted in ancient Roman art. Indeed it can be seen in an engraving of Bacchus in Bath. The *falx vinitoria* can also be seen illustrated in Medieval and Renaissance art. Examples are illustrated in medieval books of prayers and horticultural manuals, as illustrated in the next chapter (Fig. 3.2). It can be seen in the engraving of Bacchus by Crispin de Passe c. 1565-1637.

Sadly no specimen of a *falx vinitoria* has been found to date in Romano-British sites (or at least by 2001, see Brown *et al.* 2001). On the other hand an unsophisticated version of the *falx vinitoria,* a sort of bonsai billhook, has been found in many Romano-British sites concentrated to the south-east of a line extending from the Severn to the Humber, but one even as far north as Stirling (Figure 2.4).

Brown *et al.* (2001) have argued that this tool may have been used in viticulture. If this interpretation is correct then the specimen found at Stirling is either an example of the triumph of optimism over experience, or perhaps, lost loot from a Pictish raid.

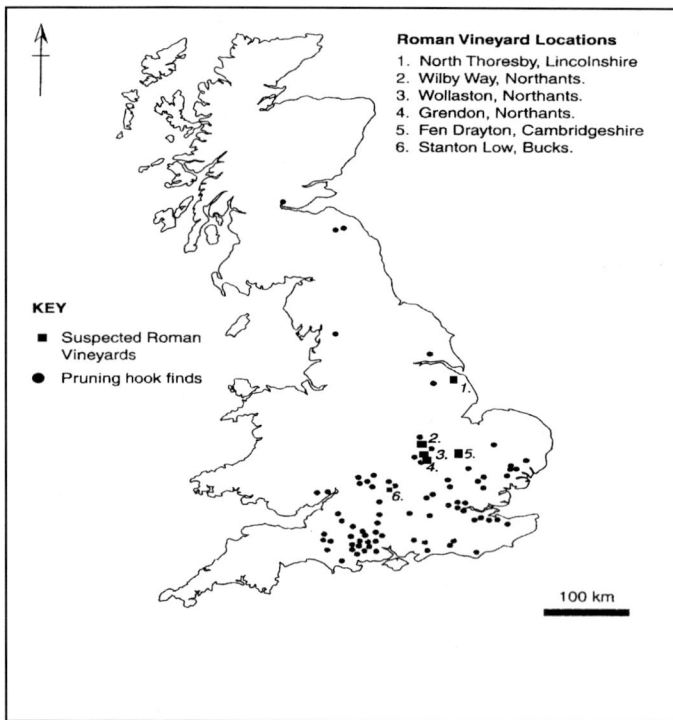

Roman Vineyard Locations
1. North Thoresby, Lincolnshire
2. Wilby Way, Northants.
3. Wollaston, Northants.
4. Grendon, Northants.
5. Fen Drayton, Cambridgeshire
6. Stanton Low, Bucks.

KEY

■ Suspected Roman Vineyards

● Pruning hook finds

100 km

Figure 2.4 Map showing where specimens of bonsai billhooks, possibly used for pruning vines, have been found left lying around in old Romano-British sites.

(From Brown et al. 2001, with permission of Antiquity Publications Ltd.).

Some 20 sites in England have been identified as Roman vineyards, many on very dubious evidence. The most convincing Roman vineyards are in the Nene Valley of Northamptonshire, where their abundance has lead to the area being compared with the present day Moselle wineland (Meadows, 1996). These will now be described as a case history of Romano-British viticulture.

2.3 THE ROMAN 'CÔTES DE NORTHANTS.'

Some 27 acres (11 hectares) of vineyards have been discovered on the flood plain of the River Nene in and around Wollaston, Northants. This area, by no means fully excavated, straddles the Roman Road from Irchester to Towcester. It is of $2^{nd} - 3^{rd}$ Century age (Jackson, 1995; Meadows, 1996; and Brown *et al.* 2001). In this region over 6km of curiously shaped trenches have been excavated to date (Fig. 2.5). The trenches are cut into river terrace gravel of the Nene valley. The trenches are some 5m apart, 0.85m wide and 0.3m deep. They are steep-sided and have flat bottoms. There are vestiges of postholes and of root balls within the trenches (Fig. 2.6). The shape of the ditches, and the pattern of stakes, is comparable to that known as 'pastinatio', described in some detail in 'De Re Rustica' written in about 66AD by Columella, the Alan Titchmarsh of Ancient Rome. This work contains extensive technical detail on agriculture in general and viticulture in particular.

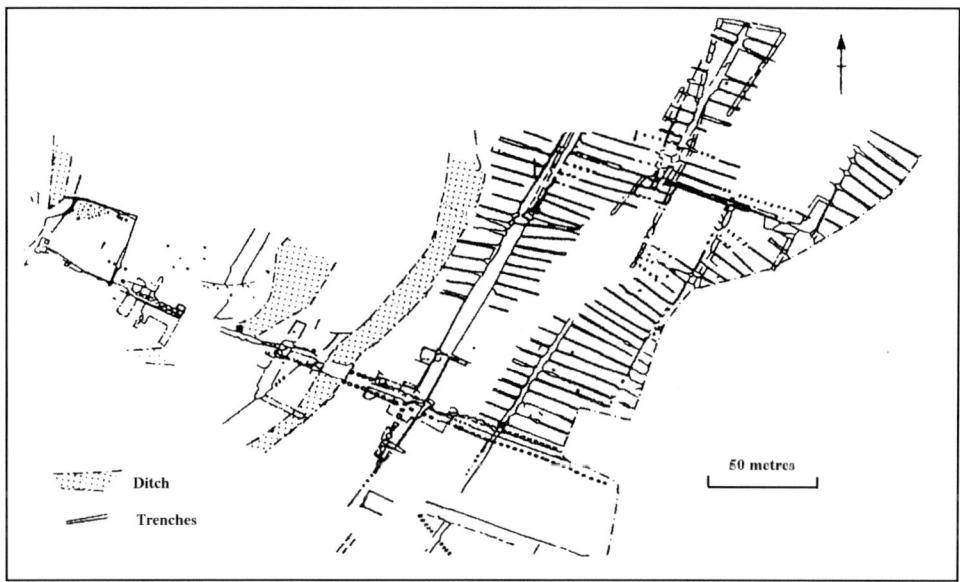

Figure 2.5 *Trenches and ditches of the Wollaston Romano-British vineyard of the Nene Valley (From Brown, et al. 2001, with permission of Antiquity Publications Ltd.).*

Support for the interpretation that this extensive system of 'pastinatio' style trenches were indeed for vines, is supported by the presence of '*Vitis*' pollen. Calculations show that the Wollaston vineyards were capable of producing over 11,000 litres of wine a year. They possibly provided *vin de pays* for the soldiers of the IX Hispana Legion, garrisoned at Lindum (Lincoln), Eburacum (York) and on Hadrian's Wall further north. Bonsai billhooks have been found in the surrounding countryside, though not in these excavations.

It is highly improbable that the archaeologists who unearthed the Nene Valley vineyards serendipitously stumbled on the only such large site in Britain. Statistically it is reasonable to assume that there must be many more Romano-British vineyards than the small number known to date. This is particularly likely when one considers that climatic conditions for viticulture improved southwards from Northamptonshire across the province.

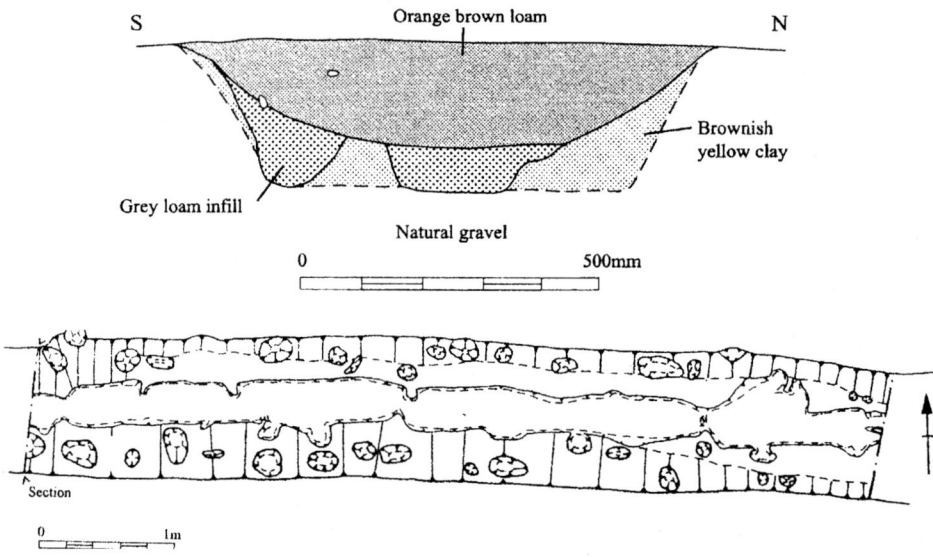

Figure 2.6 Upper: Cross-section and plan of trenches from the Wollaston Roman vineyard with imprints of root balls and post holes arranged in the 'pastinatio' style described by Columella in 66AD. (From Brown, et al. 2001, with permission of Antiquity Publications Ltd.).

Right: Photograph of the same by courtesy of I. Meadows. © Northants. County Council.

2.4 ROMAN VINEYARDS: A REVIEW OF THE EVIDENCE

Since the extent and even the mere existence of viticulture in Roman Britain were once debated it is appropriate to conclude this chapter with a review of the evidence, literary and archaeological. Domitian's edict banning viticulture north of the Alps in AD90 did not specifically mention Britain. On the other hand Probus' edict lifting the ban in AD280 specifically does mention Britain. Cynics would argue this could have been an Imperial joke, since no one in his or her right mind might consider such an enterprise.

Some 20 sites in England have been identified as Roman vineyards by people with extensive archaeological expertise and none. Roman vineyards identified on the basis of oral tradition and hillside terracing seldom survive close examination. Even those sites identified on the evidence of *Vitis* pollen and seeds do not necessarily prove that wine-making took place. The pips and pollen may have been imported from Euroland in dessert grapes or raisins. Similarly it could be argued that grape detritus, and the vineyards of the Nene Valley, merely point to a flourishing industry growing dessert grapes and drying grapes for raisins. True doubters require the excavation of a Romano-British wine press, before the matter is conclusively resolved.

The widespread distribution of the *Falx vinitoria* provides further supporting evidence for extensive viticulture. The occurrence of home made wine amphorae at several sites across southern Britain lends credence to widespread wine making. Dating evidence shows amphora production began in the 1st Century AD, very early on during the Roman occupation.

Given that the conquered Celts enjoyed wine, and that their conquerors considered wine synonymous with civilisation, it is incredible if they did not combine their physical and intellectual powers to make wine to their mutual advantage and pleasure. The idea that the Nene valley vineyards produced raisins for sale in the Lincoln branch of Marcus and Spencius is whimsical.

2.5 BIBLIOGRAPHY

The history of British viticulture in this and the two following chapters, is based to some extent, and is covered in far greater depth, by the following:

Barty-King, H. 1977. A tradition of English Wine. Oxford Illustrated Press. Oxford. 250pp.

Barty-King, H. 1989. A Taste of English Wine. Pelham Books. London. 209pp

Skelton, S. 2001. The Wines of Britain & Ireland. Faber & Faber. London. 531pp.

2.6 REFERENCES

Archaeologia Cantiana. 1856. Notes of Vineyards. 1. 232

Brown, A.G., Meadows, I., Turner, S.D. and D.J. Mattingly 2001. Roman vineyards in Britain: stratigraphic and palynological data from Wollaston in the Nene Valley, England. Antiquity. 75. 745-757.

Carr, B. 1993. In: Fortescue, S.E.D. 1993. The House on the Hill. The story of Ranmore and Denbies. Denbies Wine Estate. Denbies.125pp.

Columella, L.J.M. edited and translated by Ash, H.B. 1941. De Re Rustica. Books I – IV. Cambridge, Harvard University Press (Loeb Classical Library) 461 + xxxiii pp.

Fortescue, S.E.D. 1993. The House on the Hill. The story of Ranmore and Denbies. Denbies Wine Estate. Denbies.125pp.

Frere, S. 1978. Britannia: A History of Roman Britain. (Revised Edition) Routledge & Kegan Paul. London. 487pp.

Godwin, Sir H. 1956. The History of the British Flora. Cambridge University Press. Cambridge. 290pp.

Jackson, D. 1995. Archaeology at Grendon Quarry, Northamptonshire. Part 2. Other Prehistoric, Iron Age and later sites excavated 1974-5 and further observations between 1976-80. Northamptonshire Archaeology. 26. 3-32

Jones, G. & Legge, A. 1987. The grape (*Vitis vinifera*) in the Neolithic of Britain. Antiquity. 61. 452-455.

Meadows, I.D. 1996. Wollaston: The Nene Valley, a British Moselle? Current Archaeology. 150. 212-15.

Planchais, N. 1972-2. Apports d'analyse pollinique a la connaissance de l'extension de la vigne au Quaternaire. Naturalia Monspeliensia, Serr. Bot. 23-24. 211-223.

Sackville-West, V. 1949. Introduction. In: Hyams, E., The Grape Vine in England. The Bodley Head. London. 5 – 8.

Seely, F. & Drummond-Murray, J. 2005. Roman pottery production in the Walbrook valley. Excavations at 20-28 Moorgate, City of London, 1998-2000. Molas Monograph 25

Symonds, RP. 2003. Romano-British Amphorae. In: Amphorae in Britain and the Western Empire. Jl. Roman Pottery Studies. 10. 50-59

Tacitus, P.C. 97-98. Agricola. (published by Penguin Books. 1948. 175pp.)

Webster, D. H., Webster, H, and Petch D, 1967. A possible vineyard of the Roman-British period at North Thoresby, Lincolnshire. Lincolnshire History and Archaeology. 2. 56-61

Williams, D. 1977. A consideration of the sub-fossil remains of *Vinis vinifera* L. as evidence for viticulture in Roman Britain. Brittania. 8. 327-34.

Williams, J.H.C. 1998. Imitation or Invention? A New Coin of Tasciovanus. The Numismatic Circular. Vol. CVI. No. 8. 350-351.

Woodfield, C. 1989. A Roman site at Stanton Low, on the Great Ouse, Bucks. Arch. Jl. 135-278.

CHAPTER 3.

Viticulture in the

Medieval Warm Period

This district, too, exhibits a greater number of vineyards than any other county of England, yielding abundant crops and of a superior quality…'
WILLIAM OF MALMESBURY (1095-1143) writing of the Vale of Gloucester

3.1 THE SAXON ANGLE

With the collapse of Roman rule in Britain in the fifth century it is reasonable to suppose that vineyards and civilisation collapsed concomitantly. The incoming Angles, Jutes and Saxons refreshed themselves with ale and mead. There is evidence that the Dark Ages coincided with, and may have been initiated by, a drop in temperature. This is variously termed the Saxon Sag, or the Dark Age Drop. It is interesting to speculate on what beverage the surviving Christian communities of the Celtic fringe used for Holy Communion. No doubt when St Augustine landed to convert the Angles in 597AD, he stimulated the import of wine from mainland Europe for ecclesiastical, medicinal and convivial purposes.

There is some evidence of viticulture during the later Saxon period. In 731 the Venerable Bede, wrote in his 'History of the English Church and People' that *'vines are cultivated at various localities'*. He continued in a less probable vein that in Ireland too *'there is no lack of vines'*. There is evidence that only a few years later wine was imported from mainland Europe. Alcuin of York (c. 735 – 804) arranged for wine to be imported from France. Vandyke Price (2002) has fingered him as the first English wine writer and critic. On his return to York after a sabbatical in France Alcuin wrote *'the wine is gone from our wineskins, and bitter beer rageth in our bellies'*. It is interesting to note that he writes of wine stored in skins, as opposed to the amphorae of the Roman age. When it came to selecting skins for storing wine, apparently a goats' scrotum was the container of choice. Not many people know that, and we are grateful to Vandyke Price (ibid.) for sharing it with us.

King Alfred (871-901) legislated the compensation due for vineyard vandalism. This has been used as evidence for Wessex wine. It has been pointed out, however, that this text is recycled from Exodus Chapter 22, perhaps to add authority to Alfred's legal compendium (Unwin, 1990). Less than a century later, however, there is evidence of three vineyards in Somerset, then a shire of the Saxon Kingdom of Wessex. In 956 King Edwy (955-959) granted vineyards at Meare and Panborough to the Benedictines of adjacent Glastonbury Abbey. King Edgar (959-75) gave a vineyard at Watchet on the Bristol Channel coast to Abingdon Abbey in Oxfordshire (Hooke, 1990). It is reasonable to believe that there were more Saxon vineyards, than surviving records reveal.

3.2 FROM SAXON SQUALOR TO NORMAN 'KNOW HOW'

The arrival of the Normans after the Battle of Hastings in 1066 lead to a major change in the drinking habits of the ruling classes. Saxon mead and ale was replaced by wine as the refreshment of aspiring knights. A renaissance in monastic life had already begun during the reign of Edward the Confessor (1042-1066) and continued after the Norman Conquest. Wine was required both for communion and secular refreshment, not that the monks overindulged. The rules of the order of Saint Benedict permitted monks only one pint of wine a day. It was apparent that demand for wine could not be accommodated by imports alone. Many new vineyards were planted. The Domesday Book, compiled in 1086-7, itemised 46 vineyards across southern England from Somerset to East Anglia (Figure 3.1 and Table 3.1).

Unwin (1990) has analysed the descriptions of the Domesday vineyards in detail. Many were in the hands of the Church. The meticulous accounts given by the clerks reveal in excess of $12^{1/2}$ acres and $124^{1/2}$ arpents of vineyards. An arpent was a unit of area applied particularly to vineyards and is believed to have been approximately the same as, or slightly less than, one acre. One arpent is equivalent to 100 square poles or perches. Calculations through the highways and byways of perches, poles and rods, reveal that there were at least 136 acres (about 55 hectares) of vineyards across southern England. Despite the customary precision of the clerks, this can only be an approximation. The usual recorded size for a vineyard was two arpents, somewhat under 2 acres. The largest, North Curry, in Somerset, was 7 acres. For Lomer vineyard in Hants, no area is given, but the entry reads *'pay to the abbot 20 sesters of wine a year'*. A sester is variously given as between 24 and 32 fluid ounces. The 2 arpents of Rayleigh vineyard in Essex yielded *'20 muids of wine, if it does well'*. The muid is a fluid measure as improbable as it sounds unpleasant. One is tempted to speculate that some whimsical Saxon villein invented it to mystify and confuse his new Norman masters. Apparently though, the muid is genuine, and is equivalent to 36 gallons (164 litres). Interestingly this calculates out as a yield of 16 hl/ha, slightly less than current English yields of around 20 hl/ha. Calculations reveal that a typical Norman 2 arpent vineyard, yielding 20 muids of wine a year, would provide the wine ration for 15.78 Benedictine monks per annum, or 15.82 Bm/pa in a Leap Year, enough for a modest monastery, or the choir of a large one. The total area of vineyards recorded in the Domesday Book would satisfy the needs of 568 Bm/ per annum, or 570 Bm/pa in a leap year. Presumably much more wine was imported from France, and new vineyards continued to be planted in monasteries throughout medieval times for both sacramental and medicinal purposes.

There is much evidence in the Domesday Book that many vineyards were newly planted. A peculiar style of the text is that it records the pre- and post-conquest wealth of each property, noting the area under plough, the numbers of slaves, livestock, or beehives, and the rent, TRE (*Tempus Rex Edwardus*) and now (1086-7). Thus there is a litany of *'There was and is one plough'*. It is noteworthy that the vineyard entries are generally in the form of *'Now two arpents of vineyard'*, implying that there was not one TRE.

Further confirmation of this supposition is provided by several entries that specifically mention that the vineyards were newly planted, and had yet to bear fruit. Though the manor in which each vineyard was planted is known, the exact spot is not. Thus it is hard to know to what extent Norman viticulturalists used landscape to enhance the success of a vineyard.

The single arpent vineyard recorded on one of the three marshy islands of Muchelney, Midelney and Thorney, in the Somerset levels, does not suggest much appliance of science, since wines detest waterlogged roots. On a nearby south-facing slope at Pilton, however, the Abbot of Glastonbury planted a vineyard in the middle of the 13th century. This site was replanted in the 1960's, though has since returned to fallow.

Figure 3.1 *Map of vineyards of the Medieval Warm Period. This includes those of the 11th century Domesday Book listed in table 3.1, as well as later ones documented in table 3.2. This map is surely incomplete, since vineyards were an integral part of most monasteries and abbeys.*

By the early 12th C. William of Malmesbury was writing of the County of Gloucestershire: "This district has a greater quantity of vineyards than any other county of England, yielding abundant crops and of superior quality; nor are the wines made here by any means harsh or ungrateful to the palate, for in point of sweetness, they may bear comparison with the wines of France' (Malmesbury, c.1125). Indeed French peace treaties of the day included clauses banning the export of English wine to France (Ellis, 1833).

COUNTY	DOMESDAY NAME	MODERN NAME	SIZE	REMARKS
Beds.	Etone	Etone Socon	2 acres	
Berks.	Bistesham	Bisham	2 arpents	
	Standone	Standon	2 arpents	
Bucks.	Evreham	Iver	2 arpents	
Cambs.	Ely	Ely	3 arpents	Church land
Dorset	Deruinnestone	Durweston	2 acres	
	Odetun	Wooton Fitzpaine	2 arpents	
Essex	Ascenduna	Ashdon	1 acre	
	Belchamp	Belcamp	11 arpents	'of which one is bearing'
	Deppedana	Debden	4 arpents	'two bearing fruit, one not yet bearing'
	Hainghehan	Heddingham	6 arpents	
	Mundana	Mundon	2 arpents	
	Ragheleia	Rayleigh	6 arpents	'and it renders 20 muids of wine if it does well'
	Stanburna	Stambourne	1 arpent	AKA Toppesfield
	Stibinga	Stebbing	2.5 arpents	'… and only half is bearing.'
	Waltham	Waltham	10 arpents	'There are now 10 arpents of vineyard.'
Gloucs.	Stanhus	Stonehouse	2 arpents	
Hants.	Lammere	Lomer	?	'pay to the abbot 20 sesters of wine a year.'
Herts.	Berchamstede	Berkhamstead	2 arpents	
	Standone	Standon	2 arpents	
	Waras	Ware	4 arpents	'newly planted
Kent	Certh	Chart Sutton	3 arpents	Church land
	Cistelet	Chislet	3 arpents	Church land
	Esledes	Leeds	2 arpents	
Middx.	Coleham	Colham	1 arpent	
	Hermodesworde	Harmondsworth	1 arpent	
	Holeburne	Holborn	Not given	
	Chenetone	Kempton	8 arpents	'newly planted'
Somerset	Glastingberie	Glastonbury	3 arpents	Church land
	Nortcuru	North Curry	7 acres	
	Meare	Mere	2 arpents	Church land
	Midelenie Michelenie Torelei	Midelney Muchelney Thorney	1 arpent	'three islands'
	Wadeneberie	Panborough	3 arpents	Church land
Suffolk	Berchingas	Barking	2 arpents	Church land
	Clara	Clare	5 arpents	'Now 5 arpents of vineyard..'
	Lauen	Lavenham	1 arpent	
Wilts.	Bradeford	Bradford-on-Avon	1 arpent	
	Lacoch	Lacock	½ an acre	
	Tollard	Tollard-Royal	2 arpents	
	Wilcote	Wilcot	Not given	'…a good vineyard.'

Table 3.1 *Vineyards documented in the Domesday Book of 1086-7.*

24

Extensive written evidence shows that viticulture was widespread in both the Norman and Plantagenet periods. There are many references to payments made to 'The Keeper of the Royal Vineyards' in the 12th C. Table 3.2 lists post-Domesday Book vineyards of the Medieval Warm Period. This is surely incomplete, since most monasteries and abbeys are believed to have had vineyards. Nor were they restricted to England. Gerald de Barri, AKA Giraldus Cambrensis or 'Welsh Gerald', (1146? – 1220?) described three vineyards in southern Wales. (Brewer and Dimmock, 1861-77). The vineyards of Ely are particularly instructive. The Domesday Book (1087) lists a vineyard of 3 Arpents (say 2.5 acres). This had expanded to 6 acres by 1272. Meanwhile the Bishop of Ely established his own personal vineyard of 9 acres in 1251 (!). In the 1540's, however, the monks vineyard was converted into an orchard (Pugh, et al. 2002). The evolution of Ely vineyards encapsulates what happened nationwide, with a rapid expansion from Saxon to Norman times, and a slow decline through the reigns of the Plantagenet and Tudor monarchs.

COUNTY	LOCATION	AGE	SOURCE
Berks	Windsor Castle	12th C.	Tighe & Davis 1858, Simon 1906
Cambs.	Ely	11th-13th C.	Pugh et al. 2002
Hampshire	Winchester		Twyne & Robert of Gloucester, in Barty-King 1989
Hereford	Hereford	13th C.	Barty-King 1989
	Ledbury	13th C.	Simon 1906
Kent	Northfleet	13th C.	Barty-King 1989
	Teynam	"	"
London	Fenchurch	13th C.	Barty-King 1989
	Holborn	"	"
	Vine Street	"	"
Northants.	Rockingham	12th C.	Simon 1906
Somerset	Mere	12th C.	cited in Barty-King 1989
	Panborough	"	"
	Pilton	13th C.	Bartholme, cited in Barty-King 1989
Surrey	Croydon	13th C.	Chadwick & Phillpotts 2002
	Purley	12th C.	Simon 1906
WALES			
Pembs./Dyfd	Manorbier	12th C.	Giraldus
Glamorgan	Diwybod	"	"
Gwent	Diwybod hefyd	"	"

Table 3.2 *Some documented post-Domesday Book vineyards of the Medieval Warm Period. This is surely an underestimate, since vineyards were an integral part of most monasteries and abbeys.*

Three factors lead to the decline of Medieval British viticulture. When Henry II married Eleanor of Aquitaine in 1152, the vineyards of Bordeaux were part of her dowry. Subsequently the Black Death (1348-70) led to a dramatic collapse of agriculture, viticulture included.

A drop in temperature in the mid-fourteenth century may have been more important even than the Black Death. Temperature declined from some 0.3°C. above a 1900 benchmark to 0.2°C. below it. This decline of 0.5°C., which extended over a century, marks the end of the so-called 'Medieval Warm Period' and the start of the 'Little Ice Age'. The abandonment of vineyards in Saxony, Prussia and the Baltic states at this time has also been attributed to this drop in temperature.

There is documentary evidence, however, that vineyards were still flourishing in the late Middle Ages. The early 15th century Peterborough Salter contains a woodcut of grapes being harvested with a *falx vinitoria*, identical to those of Roman times. A similar sixteenth century example from a prayer book is illustrated in Figure 3.2. William Camden writing in 1586 recycled William of Malmesbury's praise of Gloucestershire's vineyards as being still apposite.

Figure 3.2 *Illustration of viticulture (left) and vinification (right) in a 16[th]C. Prayer book. Note the right hand peasant in the left hand picture wielding a bonsai billhook. This may be a folk-memory of earlier happier times, or an indication that winemaking continued into the Little Ice Age © Mary Evans Picture Library.*

Enthusiasts will use these instances to demonstrate the continuity of viticulture beyond the Black Death and well into the Little Ice Age. Cynics may argue that these drawings were recycled from earlier illustrations, comparable to the model crocodiles made by the Tibu of the Tibesti Mountains of Chad, though the Saharan lakes and their fauna dehydrated centuries ago. Believe what you will. The 13th century vineyard of the Palace of the Archbishop of Canterbury, at Croydon in Surrey, may serve as an illustration of a late medieval vineyard.

3.2 CASE HISTORY: THE VINEYARD OF THE ARCHBISHOP OF CANTERBURY, CROYDON, SURREY

The selection of a case history with which to illustrate a Medieval Warm Period vineyard is difficult. Few are known from both written and archaeological evidence. One such, however, is the Palace of Croydon, Surrey, owned by the Archbishops of Canterbury from the 8th century until the middle of the 17th century. Lambeth Palace library contains documents that describe a vineyard in the palace grounds.

There is a specific mention of cultivation and the replanting of an existing vineyard in 1236-7 (Lambeth Palace Library: ED 1193). The Croydon Palace was excavated during 1998-9 (Chadwick & Phillpotts, 2002). The vineyard was located on a well-drained site of Thanet Sands (Palaeocene). The excavations revealed that the vineyard measured some 30 x 20m and was enclosed by a wall, whose height it is today impossible to know (Fig. 3.3).

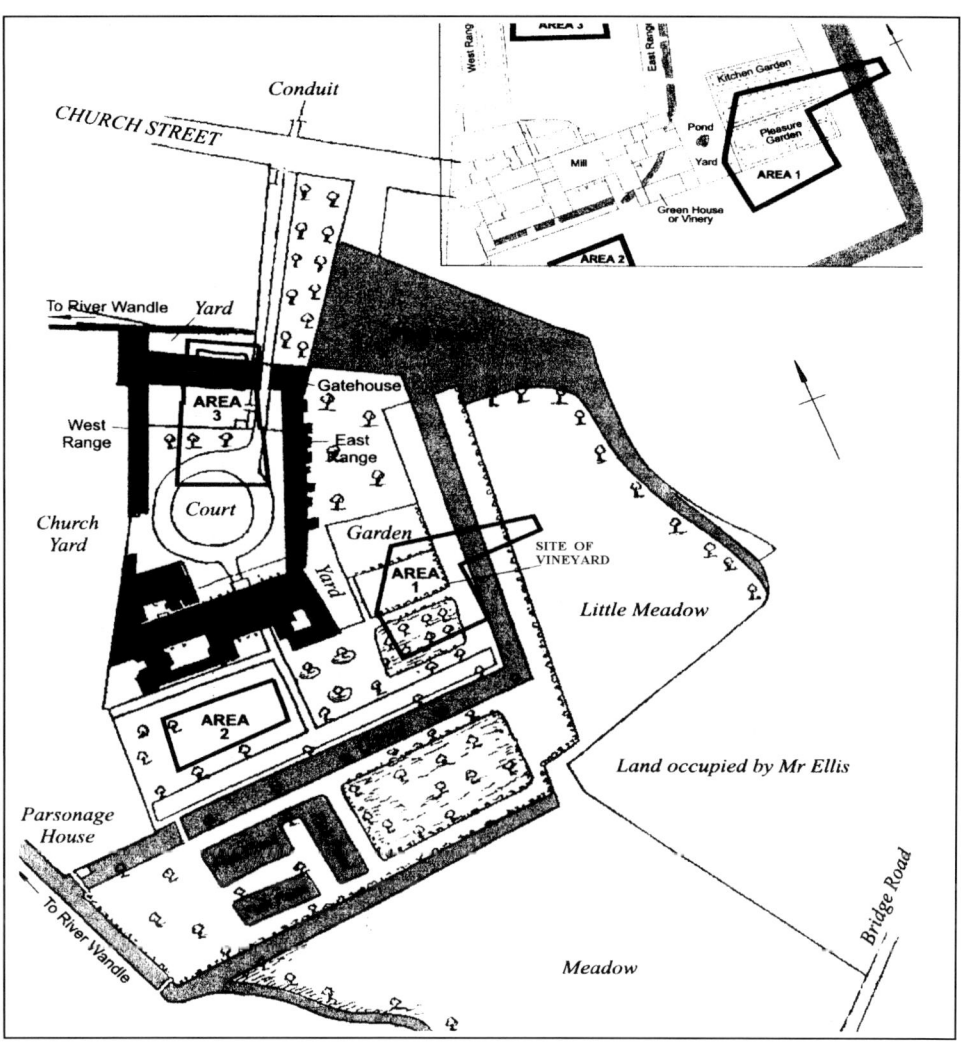

Figure 3.3 *Plan of Croydon Palace showing the location of the 13th century vineyard. From Chadwick & Phillpotts, (2002) © Wessex Archaeology. The small size (30 x 20 m) and way in which the vineyard was an integral part of the estate grounds foreshadows the style of vineyard of the 'Little Ice Age' described in the next chapter.*

Though detailed botanical studies of the site were carried out, neither pips nor pollen of *Vitis* were recorded. This, in itself, is valuable information. The absence of *Vitis* pollen and pips in a known vineyard indicates that they must have a low preservation potential. Thus no conclusion should be drawn if they are not found in suspected ancient vineyards.

Assuming a yield of 16hl/ha of wine, the figure calculated for 11th century vineyards of the Domesday Book, the Archbishop's vineyard at Croydon would have produced about 1000 litres of wine a year. This would have sufficed for 4.629 Benedictine monks per annum, 4.642 Bm/pa in a Leap Year. The Croydon Palace vineyard is comparable in size to the one in the Little Park of Windsor Castle 80 years earlier (Tighe and Davis, 1858).

It is interesting to speculate if the Croydon Palace 13th century vineyard was just an archiepiscopal showpiece, and that the Archbishop had more substantial vineyards on his ecclesiastical manors, as in the 11th century. Perhaps, however, its small size indicates the change from vineyards being run as commercial ventures to becoming the hobby of country gentry, as in the Little Ice Age described in the next chapter.

3.3 REFERENCES

Bede, the Venerable, 731. A History of the English Church and People. Penguin Classics. London.341pp.

Brewer and Dimmock, 1861-77. The Works of Giraldus Cambrensis. 7 Vols.

Camden, W. 1586. Britannia, or a Chorographical Description of Great Britain and Ireland (In Latin). English translation by P Holland 1610.

Chadwick, A.M. & Phillpotts, C. 2002. The Archbishop's 'great stable': excavations and historical research at the Old Palace School, Croydon, 1999. Surrey Archaeological Collections. 89. 27 - 52

Ellis, H.K. 1833. A General Introduction to the Domesday Book.. 2 Vols. London. Commissioners of Public Records.

Hooke, D. 1990. A Note on the Evidence for Vineyards and Orchards in Anglo-Saxon England. Jl. Wine Research. 1.1. 77-80.

Lambeth Palace Library: ED 1193.

Malmesbury, W. c.1125. Gesta Regum Anglorum.

Pugh, RB (Ed.) 2002. A History of the County of Cambridge and the Isle of Ely. Vol. 4. 33-40

Simon, AL. 1906. The History of the Wine Trade in England. Vol. 1. Holland Press. London

Tighe & Davis 1858. Annals of Windsor. Vol. 1. Unwins. London.

Unwin, T. 1990. Saxon and Early Norman Viticulture in England. Jl. Wine Research. 1. 1. 61-76

Vandyke Price, P., 2002. Curiosities of Wine. Sutton Publishing. Stroud.196pp.

CHAPTER 4.

Viticulture in the

Little Ice Age.

...produced most excellent good wines, and a very great quantity of them.
DEFOE writing of Deepdene vineyard 1724-6.

4.1 GLOBAL COOLING: A MEASURED VIEW

As noted at the end of the previous chapter, viticulture began to decline through the 14th, 15th and 16th centuries. There is no doubt that this decline was partially due to socio-political factors, such as access to cheap European imports, the Black Death (1348-70) and the dissolution of the monasteries by Henry VIII in 1536-9, with the concomitant loss of their vineyards. These events were taking place, however during a period when it is estimated that the mean temperature of Britain dropped by nearly 1^{0}C from a maximum at about 1200, to a nadir at around 1450, from which it failed to recover, apart from one or two minor oscillations, until the start of the Industrial Revolution in the middle of the 19th century.

The evidence for climate change before the advent of accurate measuring methods needs to be reviewed. The thermometer was invented by Ctesibius of Alexandria in the 3rd century BC, and dubiously reinvented by Galileo. By the 1670's accurate measurements were being made of English ambient temperature by men such as Robert Hooke and John Locke. There were, however, other rough and ready methods of guestimating temperature. Rural parsons, such as Gilbert White of Selbourne, Hants. (1720-93) and Parson Woodford, of Norfolk (1740-1803) kept journals that included descriptions of weather. White, in particular, took detailed observations of natural history in general and weather in particular. Many of these observations were phenological, noting the advent and departure of migratory birds, the date when Timothy, his tortoise, entered and emerged from hibernation, the blossoming of various flowers, and the start of harvest. During the bitterly cold winters of this period the onset of spring was marked by the pitter patter of tiny feet as birds melted and dropped from the trees to which they had been frozen all winter. Notes were taken of those nights when the contents of chamber pots were found frozen in the morning, or, as White politely recorded, 'tonight it did freeze under my bed'. This is a method of thermometry endorsed neither by the British Standards Institute nor the European Union. Phenological data have to be interpreted very carefully (Lamb, 1964). The decline in the frequency of the River Thames freezing over, for example, had as much to do with the liberation of tidal currents when Old London Bridge was removed in the mid 18th century, as to do with global warming. Subsequently far more scientific palaeothermometers have developed that enable past temperatures to be measured with great accuracy.

These include dendrochronology, measurements made from tree rings, and oxygen isotope ratios measured from cores recovered from ocean floor sediments and from the polar ice caps. These enable temperature changes to be measured and climate change determined far back into prehistoric time. Suffice to say that the temperature drop of the Little Ice Age is well proven from both contemporaneous records and subsequent palaeothermometric analysis. There is, however, abundant written evidence of vineyards across southern England throughout the Little Ice Age of the 15th to 19th centuries. Several books of the period on viticulture, include 'An Excellent Way for the Planting of Vines According to the French and German Manner and Long Practiced in England', by William Hughes in 1670. William Cobbett (1763-1835), though celebrated for his 'Rural Rides', also wrote a lesser-known work 'The English Gardener' (1829). In this book Chapter 286 describes viticulture. This is several pages in length and accompanied by illustrations, from both of which it is apparent that the author had first hand experience of viticulture. John Rose's 'The English Vineyard Vindicated' (1666) does have a defensive-sounding title, however, and suggests that viticulture was challenging. It is Barty-King's view (1977) that the decline in viticulture during the Little Ice Age was more imagined than real, and had little to do with climate change. He attributes the low level of viticulture practiced in Britain to 'the sloth of the inhabitants, and the indisposition of the climate' (ibid., p. 60). Table 4.1 provides details of some of the known vineyards of the Little Ice Age, and Figure 4.1 shows their distribution.

COUNTY	NAME	DATE	SIZE	SOURCE
Derbyshire	Darley Abbey	1557	Unknown	S. Pegge, 1771
Hants. (Isle of Wight)	St Lawrence	1792-	3 acres	J & J Jones, 1987
Herts.	Hatfield	1611-1661?	4 acres	S.Pepys, 1611 & 1667
Kent	Boughton Malherbe	1580	Unknown	Barty-King, 1989
	Egerton	1580	Unknown	Barty-King, 1989
	Goddington	c. 1650	Unknown	Tod, 1910
	Great Charte	1620	Unknown	Hartlib, 1659
	Harrietsham	1580	Unknown	Barty-King, 1989
London	Chelsea	c. 1790	Unknown	Skelton, 2001
	St James's	1615	Unknown	Barty-King, 1977
Surrey	Albury Park	1662- ?	Unknown	J. Evelyn
	Deepdene	1655-1725	7 acres	Aubrey, Evelyn & Defoe
	Godalming	1720-85	Unknown	Barty-King, 1977
	Oatlands	1615	Unknown	Barty-King, 1977
	Painshill Park	1740-1790	13 acres	Cobbett, 1829 & others
	Wimbledon	1786	Unknown	Skelton, 2001
	Writlemarsh	1665		S.Pepys, 1 May 1665
Sussex	Arundel	1772-1804	Unknown	Tithe maps

Table 4.1 *Vineyards of the 'Little Ice Age' from the 16th – 19th centuries.*

Figure 4.1 *Map of Little Ice Age vineyards compiled from data in Table 4.1. Note how, apart from one tough northerner, they were restricted to the south east of England, unlike in former and later times.*

Not surprisingly they are mainly in the south and east of England. One outlying exception was Darley Abbey in Derbyshire in the mid-16th century. In his 'The English Gardener' Cobbett remarks that though vines are now (1829) grown largely as curiosities in sheltered gardens, vineyards once flourished as far north as Warwickshire, and produced wine as good as France. It is apparent that, unlike the Medieval and Industrial warm periods that preceded and followed the Little Ice Age, vineyards were generally small.

31

An exception to this general rule may be the vineyard at Hatfield House, Hertfordshire, planted by Lord Salisbury in 1610. 30,000 French vines were planted on a 4-Acre site. Samuel Pepys visited this vineyard in 1661 and 1667. There are records that it was still in existence in 1924. Reliable information is scarce, but it appears that most vineyards of the 'Little Ice Age' were considerably smaller. Gentlemen of leisure, many of whom were inspired by the Grand Tour, planted vineyards as curiosities. Estates of this period were replete with follies, belvederes, lakes, waterfalls, hermitages, grottoes, miradors, caves and vineyards. Their owners sought to reproduce the illusion of Mediterranean classical grandeur and climate. Nonetheless, though vineyards were small, writers describe English wine of the period as comparable in quality to imported wines.

Cobbett (1829) is one of several contemporaries who left written accounts of the noted Little Ice Age vineyard of Painshill Park, near Cobham, Surrey. The Honourable Charles Hamilton acquired this estate of some 250 acres with borrowed money in 1738. He developed the park into ornamental pleasure grounds, complete with an artificial lake, waterfalls, water wheel, grotto, mausoleum, Turkish tent, Chinese bridge, Gothic temple, Gothic tower, ruined abbey, hermitage, and vineyard, all of which have undergone meticulous modern restoration (Fig. 4.2).

Figure 4.2 *Painshill Park vineyard, Cobham, Surrey. Courtesy of the Painshill Park Trust. Painshill vineyard was first planted by the Hon. Charles Hamilton in 1748, during the Little Ice Age, and replanted at the end of the last century.*

The vineyard was more successful than the hermitage. A man engaged to star as the hermit was contracted to live in the hermitage for 7 years, at the end of which period he would receive £700. Legend has it that he was found in the local alehouse tired and emotional after only 3 weeks. The vineyard fared much better. According to contemporary accounts it covered some 10 acres (4 hectares), part on the flat ground to the north of the Portsmouth road, part to the south. The southern part of the vineyard was cunningly planted on a well-drained southerly slope of Bagshot Sand (Eocene), at the foot of which the lake provided additional radiation reflected from the rays of the sun (Fig. 4.3).

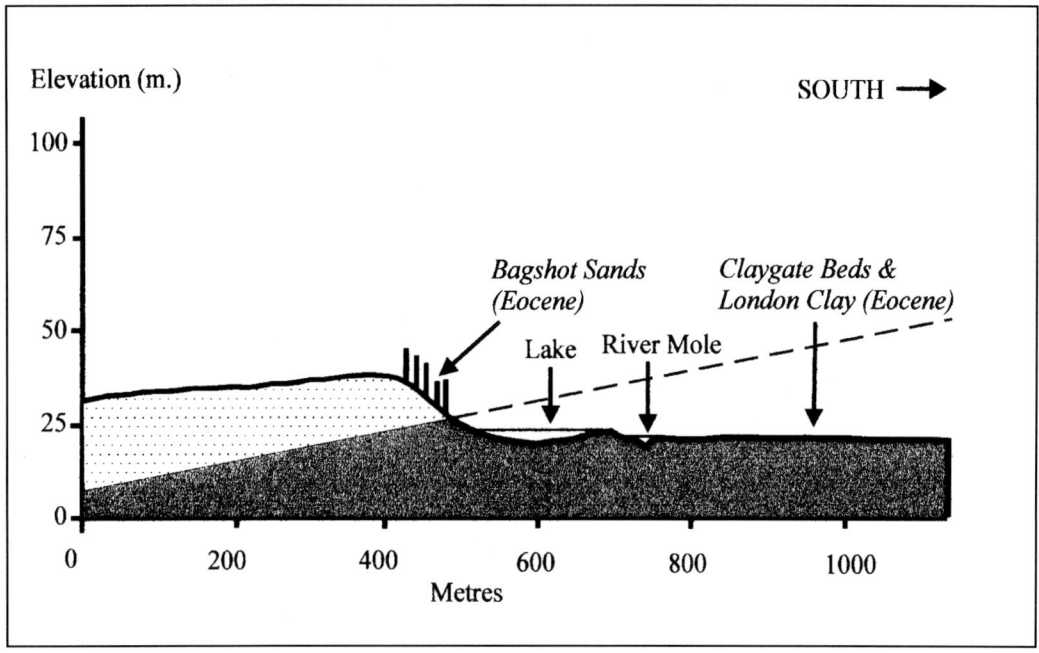

Figure 4.3 *Geological cross-section of Painshill Park vineyard. It is located on a well-drained sunny south-facing escarpment of Bagshot Sands overlying impermeable Claygate beds and London Clay (all of Eocene age). Solar radiation is enhanced by reflection from the artificial lake at the foot of the slope. This is filled by water from the adjacent River Mole brought up by means of a waterwheel.*

The Painshill Park vineyard was planted with two types of Burgundy grape, Auvernat (Pinot Noir) and Miller (Pinot Meunier). Both red and white sparkling wines were produced. The latter apparently deceived the French Ambassador, Le Duc de Mirepoix, into believing it to be Champagne, but then he was a diplomat. Painshill Park vineyard flourished for over 40 years, until 1790. The southern part of the vineyard has now been replanted (1992 – 4) with Chardonnay, Pinot Noir and Seyval Blanc; varieties as close as possible to the originals. Some 10 miles to the south of Painshill lies abandoned one of the best-documented and longest-lived vineyards of the Little Ice Age. This will now be described in detail.

4.1 CASE HISTORY: DEEPDENE VINEYARD, DORKING.

One of the most celebrated vineyards of the Little Ice Age was at Deepdene, Dorking, in Surrey (Mercer, 1977; Mercer and Jackson, 1996). Deepdene vineyard was planted by Charles Howard, fourth son of Henry Frederick, Earl of Arundel. Charles Howard, a somewhat eccentric gentleman, set himself up as a solitary hermit, complete with gardens, caves, a chapel and an elaboratory (*sic*). His property was described in considerable detail, with a mixture of wonderment and perplexity by several visitors. John Evelyn, diarist, horticulturalist and neighbour, describes several visits to Deepdene between 1655 – 70, remarking on '*divers strange plants*'.

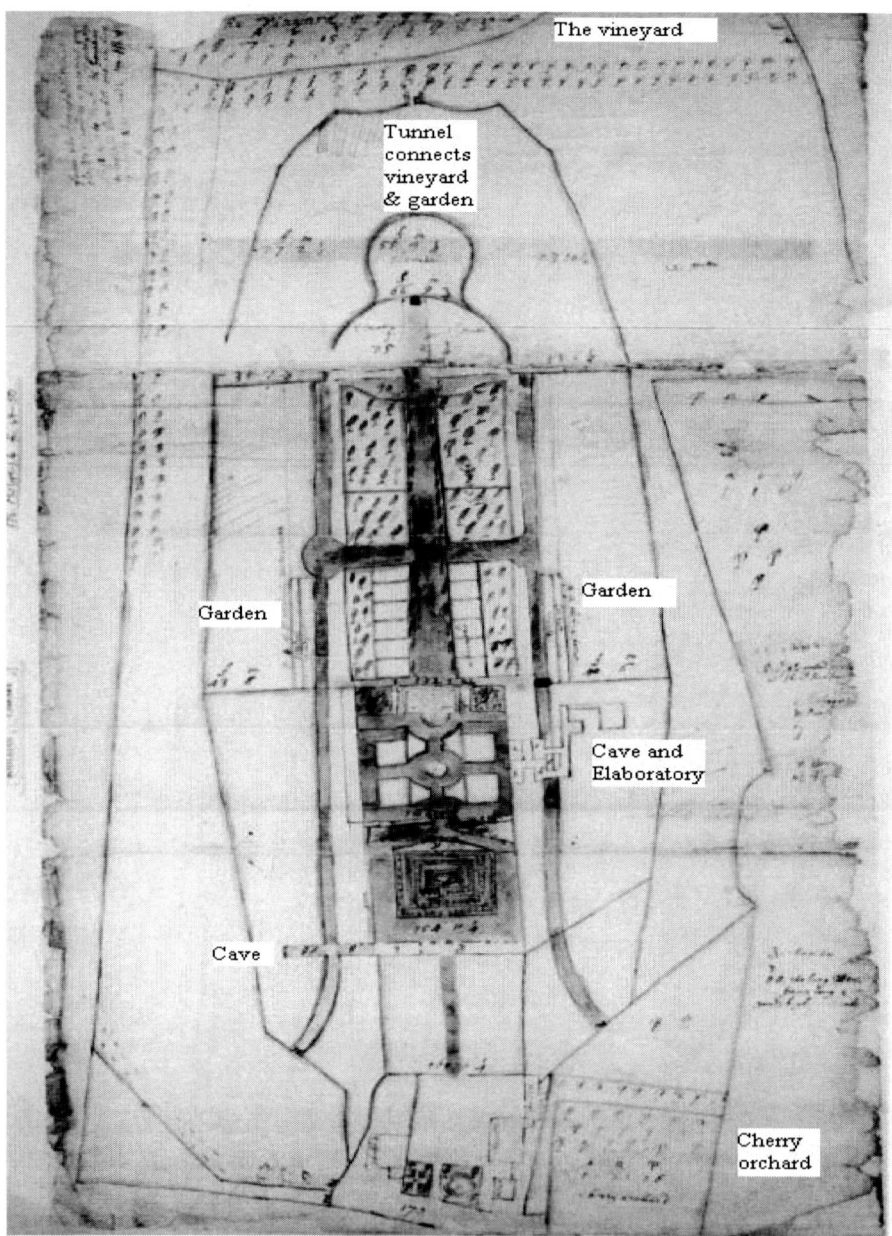

Figure 4.4 *Plan of Charles Howard's estate and vineyard at Deepdene, Dorking, Surrey c. 1673 (From the Bodleian Library, Oxford MS Aubrey 4, ff.49-50. Reproduced with permission). Top of page is south-east. The vineyard of over 7 acres (3 hectares) lay to the south east of the gardens.*

John Aubrey, famous for his 'Brief Lives', visited in 1673, and produced a plan of the estate (Figure 4.4) showing that the vineyard covered 7 acres, 1 rood and 1 pole (2.9 Hectares). This is substantially larger than many modern commercial vineyards. A yield comparable to those of medieval vineyards (16 hl/ha) would have produced some 4,686 litres of wine per annum. A yield comparable to those of modern vineyards (22 hl/ha) would have produced some 6,440 litres of wine per annum. These would have satisfied the wine ration of 22.59 and 31 Benedictine monks/per annum respectively, enough for a modest monastery. Charles Howard must have been a very happy man. Figure 4.5 shows how the vineyard was cunningly sited on a well-drained south-facing slope of the Lower Greensand. The tunnel connecting the vineyard would have allowed easy transport of the grapes, and also cellarage (Figure 4.6), some of which exists to this day.

Figure 4.5 *View of the south-facing Lower Greensand sunny scarp on which flourished the 17th – 18th century vineyard of Deepdene. The entrance of the tunnel from the vineyard to the Elaboratory (see Fig. 4.4 on the facing page) is unknown. Like most of the abandoned 'Little Ice Age' vineyards of the Surrey Hills, this one has not been replanted by vines, but by trees. Now it only produces a meagre crop of golf balls from Dorking Golf Course at its foot.*

Daniel Defoe provides the last account of the Deepdene vineyard in production. In his 'Tour Through The Whole Island of Great Britain' (1724-6) he remarked on the tunnel connecting the vineyard to the estate, and that the vineyard *'has produced most excellent good wines and a very great quantity of them.'* John Senex's map of Surrey (1729) marks 'Dibden', and a nearby 'Vintage yard' and a 'Banqueting house'.

This would seem to indicate that even if the vineyard was abandoned then there were some vestigial traces of conviviality. Roque's Map of Surrey (1762) shows no sign of Howard's vineyard or associated buildings. It does, however, indicate adjacent properties named 'Vineyard Lodge' and 'Vintage Farm'. Edwards map of 1787 also shows 'Vineyard Lodge'. The accompanying text states *'The hanging-hill to the south of Dibden (sic) was formally a vineyard which is now destroyed.'* These contemporaneous accounts show that Deepdene vineyard was no gentleman's whim. It was planted in or earlier than 1655, and flourished up until at least 1726, a span of at least 71 years. Early 19th century maps, and subsequent ones, show no trace of Deepdene's past viticultural glories.

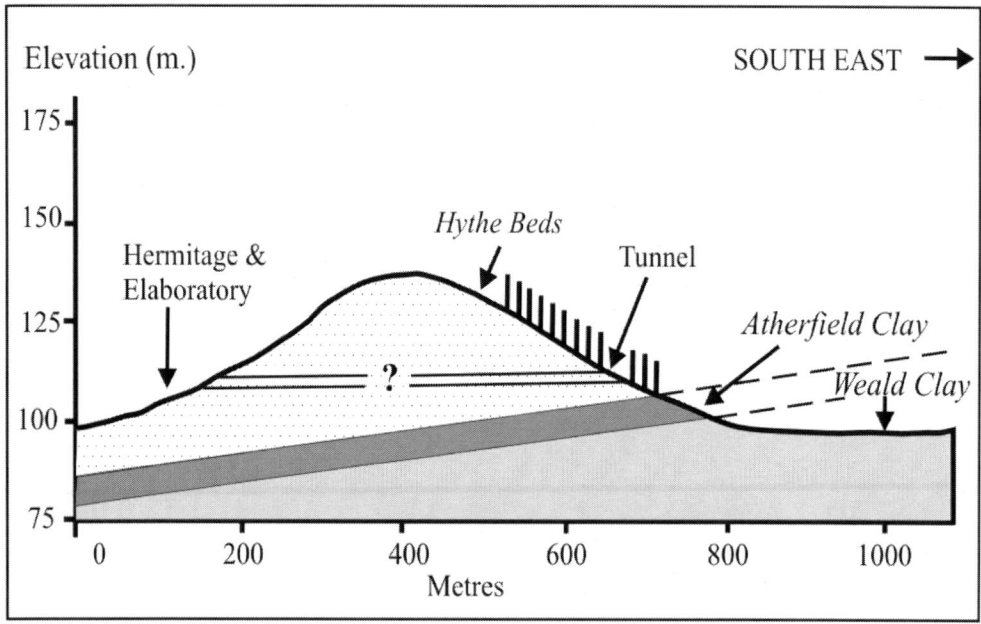

Figure 4.6 *Geological cross-section of Deepdene estate, showing how the vineyard was planted on a well-drained south-facing slope of Greensand, and connected to the rest of the property, by a tunnel.*

The Deepdene vineyard was not an isolated example, but was one of a series of vineyards planted along the south-facing scarp of the Lower Greensand in the Surrey Hills. Another member of the Howard family owned Albury Park, an estate some 10 kilometres west of Deepdene. Here in or around 1662 John Evelyn arranged irrigation for a fountain and vineyard planted on a terraced south-facing slope of Greensand (Walmsley, 1986). William Cobbett in his 'Rural Rides' (1830) describes the gardens of Albury Park in great detail, though without mention of a vineyard. Presumably it had fallen into desuetude by then.

10 kilometres to the west of the Albury Park estate James Oglethorpe MP established a third vineyard between 1720 and '32. This was also planted on a terraced southerly slope of Lower Greensand at Westbrook Place, Godalming. Oglethorpe fed his guests (French) snails fattened on his own vine leaves. The snails were washed down by his own wine. What sweeter revenge could any gardener desire?

Deepdene and the other vineyards of the Greensand slopes perhaps typify British viticulture during the Little Ice Age. Though wine was not produced on a commercial scale, a vineyard was considered as integral a part of a country gentleman's estate, as barbeques, patios, decking and water features are today.

It is interesting to note that some of the planters of these vineyards were Roman Catholic gentry, who at that period were politically disenfranchised. They appear to have chosen to retire to their country estates to practice viticulture and drink their liquid assets.

4.2 REFERENCES

Anon, 2000. A Guide to Painshill. The Painshill Trust. 40pp.

Anon. (undated) Painshill Scenes. The Painshill Trust. 21pp.

Aubrey, J. 1718. The Natural History and Antiquities of Surrey (Reprinted by Kohler & Coombes, 1975. 3. 211pp.)

Barty-King, H. 1977. A Tradition of English Wine. Oxford Illustrated Press. 250pp.

Cobbett, W. 1829. The English Gardener. Oxford Paperback Edn.1980. 335pp.

Cobbett, W. 1830. Rural Rides. Penguin English Library Edn.1967. 533pp.

Defoe, D. 1724-26. A Tour Through the whole Island of Great Britain. Penguin English Library Edn., 1971. 730pp.

Edwards, J. 1787. A companion from London to Brightelmstone (Map and text)

Hartlib, S. 1659. The Compleat Husband-man: or, a discourse on the whole art of husbandry both Forraign and Domestick Wherein many rare and most hidden secrets, and experiments are laid open to the view of all, for the enriching of these nations. London.

Jone, D. and Jones, J. 1987. The Isle of Wight. An Illustrated History. The Dovecote Press. Wimborne. 160pp.

Lamb, H. H. 1964. The English Climate (2nd. Edn.) English Universities Press, London. 212pp.

Mercer, D. 1977. The Deepdene, Dorking – Rise and Decline Through Six Centuries. Surrey Archaeological Collections. 71. 111-138

Mercer, D. & Jackson, A. A.1996. The Deepdene, Dorking. Dorking Local History Group. Dorking. 76pp.

Roque, 1762. A Map of Surrey.

Senex, J. 1729. A Map of Surrey.

Tod, H.M. 1910. Vine-Growing in England. Chatto & Windus. London.

Walmsley, R. C. 1986. A description of the Mansion and Grounds at Albury Park, Guildford, Surrey, and of the Old Parish Church. Printed privately. 16pp.

PART 2.

BRITISH WINELANDS:
PRESENT

THE WINELANDS OF BRITAIN

CHAPTER 5.

Geological controls on viticulture

Civilization occurs by geological consent – subject to change without notice.
WILL DURANT (1885-1981)

Civilization is based on wealth created either by farming the land, or by extracting rocks and fluids from beneath it. No wonder that geology, the study of the earth, is such an important subject in schools. No wonder that geologists are paid footballers salaries, fêted like pop stars, and showered with honours by grateful governments. If only. We inhabit this planet by courtesy of its geology. For example, the prosperity enjoyed by the United Kingdom for the last 30 years has been underwritten by the invisible wealth created by geologists who explored for petroleum in the North Sea.

5.1 GEOLOGICAL CONTROLS ON VINEYARD TERRAIN

Geology is one of the four variables that control the quality and character of wine. The others are climate, recipe, and grape variety. There is a huge literature on the last three of these variables, rather less on the first. Geologists have studied the relationship between geology and viticulture, termed geoviticulture, for many years, with all the collateral conviviality that such demanding research necessarily entails. The genesis of this discipline was marked by the opening address of Maurice Lugeon, eminent French Professor of Geology, at the International Congress of Wine Producers at Lausanne in 1935. For Anglophones, however, Wallace's paper at the International Geological Congress in Montreal in 1972 was the wake up call. Subsequently many books and papers have been published on the geology of wine, most notably in France, naturally (Pomerol, 1989; Wilson, 1998), but also as far afield as Slovakia (Bezak, et al. 2002), Australia (Hancock & Huggett 2004), South Africa (Bargman, 2005), the USA (Gillerman et al. 2006) and Canada (MacQueen & Meinert, 2006). Early publications on geology and wine did little more than describe the geology beneath a vineyard, with little analysis of their relationship. More recently, however, detailed analysis of the interface of geology and vines has developed. English geologists such as Dr Jeremy Leggett and the late Sir John Knill planted vineyards at Melton Lodge, Suffolk, and Shaw-cum-Donnington, Berkshire, respectively; while Denbies, England's largest vineyard at over 100 hectares, was planted on the recommendation of a notorious local geologist.

Vines grow on rocks of all types and of all ages, from the Precambrian granites to modern gravels. Superficially this suggests that geology has nothing to do with viticulture (Maltman, 2003). Indeed, when an Australian wine maker was asked what effect geology had on his product his reply was unprintable, but could be translated into wine-speak thus: *'A bouquet like a kangaroo's armpit, fruity, with a high acidity, and long tannin after burn'.*

Though vines grow on all rock types, geology controls topography and soil. These will now be considered in turn. Rocks are of three genetic types: igneous (formed from cooling magma), sedimentary, (formed by the breakdown of pre-existing rock), and metamorphic (formed by the action of heat and pressure on pre-existing rock) (Figure 5.1 and Table 5.1).

Figure 5.1 Geophantasmogram of 'The Rock's Display'd' illustrating the genesis of igneous, metamorphic and sedimentary rocks. From Wilson, in Read, 1944. Reproduced by permission of The Geologists' Association from Proceedings of The Geologists' Association, H.H. Read, Meditations on Granite. Part 2, LV, 45-93 © 1944 The Geologists' Association.

Rocks of these three origins have a wide range of minerals, and a wide range of physical properties, notably hardness, porosity and permeability. Porosity is the void space of a rock. Permeability is a measure of the interconnectivity of the pores. Permeability allows fluids to migrate through rock (Fig. 5.2), and enables rock and soil to drain. Igneous and metamorphic rocks generally lack porosity and permeability. This is because they are both normally composed of a mosaic of interlocking crystals that formed at high temperatures and pressures. Exceptionally, however, they may be fractured, and the fractures may impart permeability and some porosity. Sediment, on the other hand,

CLASS	ORIGIN	EXAMPLES
SEDIMENTARY	Deposition of particles of older rocks.	Lithified gravel, sand & mud, termed conglomerate, sandstone & shale respectively.
	Chemical precipitation & replacement	Limestone, dolomite & evaporates
METAMORPHIC	Heated by igneous intrusions (Thermal metamorphism). OR	Hornfels, granulite
	Subjected to high temperature & pressure by deep burial (Regional metamorphism)	Slate, schist & gneiss
IGNEOUS	Shallow fast cooling in volcanic eruptions	Basalt, andesite & rhyolite.
	Slow deep cooling of magma	Granite, diorite & gabbro.

Table 5.1 *Rocks, their origins and common examples.*

is deposited with space between the constituent particles. Thus sedimentary deposits, of their very nature, tend to be porous and permeable. When sediments are buried, however, minerals may precipitate out of pore fluids to cement the soft sediment and turn it into solid rock. Igneous, sedimentary and metamorphic rocks may crop out at the Earth's surface in cliffs, deserts, or on exposed mountaintops.

Landscape results from the interplay of geology and climate. Hard rocks are resistant to weathering and erosion. They thus commonly form hills. Soft rocks are more easily weathered and eroded, thus forming valleys. As a general rule igneous and metamorphic rocks are harder than sedimentary rocks, thus they form the high ground, and sediments the low ground. All geological generalisations are dangerous, however, including this one. Sedimentary strata commonly show vertical variations in hardness, and hence in resistance to weathering, creating escarpments and dip slopes separated by intervening valleys. Geology controls the altitude, orientation, topography and composition of landscape.

Figure 5.2 *Photomicrograph of a sandstone showing constituent sand grains and intergranular porosity. The interconnected nature of the pores between the sand grains make this rock permeable and easily drained.*

In some sandstones the pores are full of mineral cement, rendering the rock impermeable, unless subsequently fractured.

Climate controls the erosional processes that operate on landscape. Glacial, fluvial and desert processes all carve landscapes into distinctive forms. Consider for example the rugged terrain of Scotland and Wales, where glacial processes during the last Ice Age carved the mountains into jagged peaks, with corries, arêtes, and other glacial landforms. Contrast this with the gentle mammillated topography of southern England, which suffered only periglacial tundra conditions.

Temperature decreases at a rate of 0.6°C per 100m. Thus in hot climates vines flourish in the mountains, whereas towards the polar limits of viticulture they are cultivated at progressively lower altitudes. This point is demonstrated by comparing the elevation of the vineyards grown on the chalk in the Champagne of France, and of the English Downs, discussed later. In warm climates vines will thrive on horizontal land. Sunshine is adequate to ripen grapes. Moving towards the poles, however, the angle of the sun gradually decreases, temperature decreases, and the amount of solar radiation received by the vines correspondingly declines. There are therefore advantages to planting vineyards on north-facing slopes in the southern hemisphere, and south-facing slopes in the northern hemisphere. This is illustrated in figure 5.3 and may be expressed mathematically as:

$$I = k \sin (\alpha + \beta)$$

Where I = the intensity of radiation received on the slope
   k is a constant
   α = the angular elevation of the sun
   β = the angle of inclination of the slope to the horizontal along a meridian, i.e. to the south, in the northern hemisphere, to the north in the southern hemisphere.

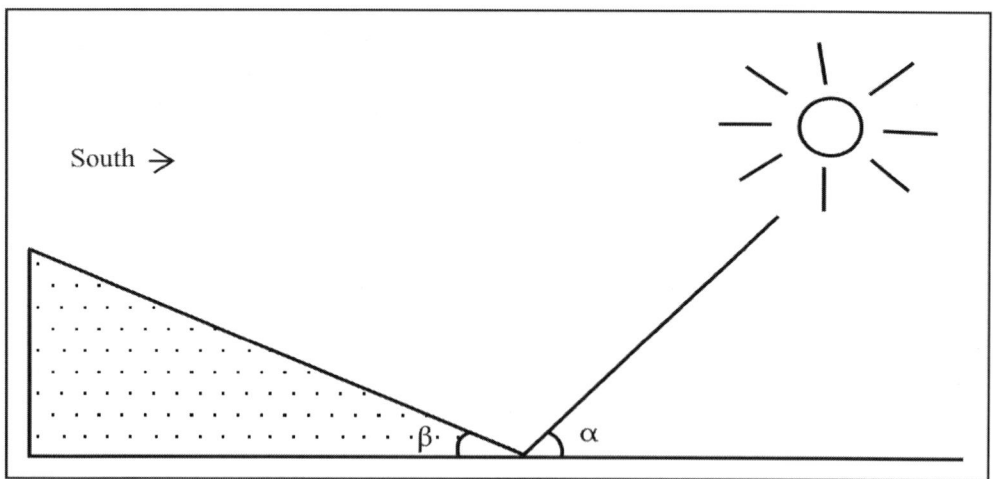

South →

Figure 5.3 Diagram to illustrate the relationship between slope and solar radiation (after Smart, 1999 and Hancock, 1999)

In southern England, where the angle of the sun is 62° at noon on mid-summer's day a vineyard on a 30° slope will receive 8% more sunlight than a vineyard on level ground. This does not sound too significant. By October, however, during the final days of ripening before harvest, the difference is 30%.

It is further held that the optimum orientation of a slope is to the east of south, since the vines then benefit from the early morning sun when the diurnal temperature is at its lowest. Easterly slopes may also be sheltered from prevailing westerly winds. For the southern hemisphere, needless to say, it is northerly slopes that are preferred to southerly ones. Vineyards planted on slopes benefit from good soil drainage in regions of high rainfall. Vineyards are best planted halfway up a slope in what is referred to as the 'thermal zone'. If too near the hillcrest the vines may be damaged by cold winds, if too close to the bottom of the slope the vines may suffer from cold air trapped at the bottom. Frost hollows are particularly dangerous, where cold air is trapped at the bottom of valleys. Late spring frosts are especially hazardous at the time of budburst, since they may not only destroy the buds, but also even large numbers of vines. Prominent headlands may provide shelter from prevailing winds. Thus, as the cross-section in Figure 5.4 shows, vineyards ancient and modern, have been have been planted in southern England to take advantage of geology and the resultant landscape.

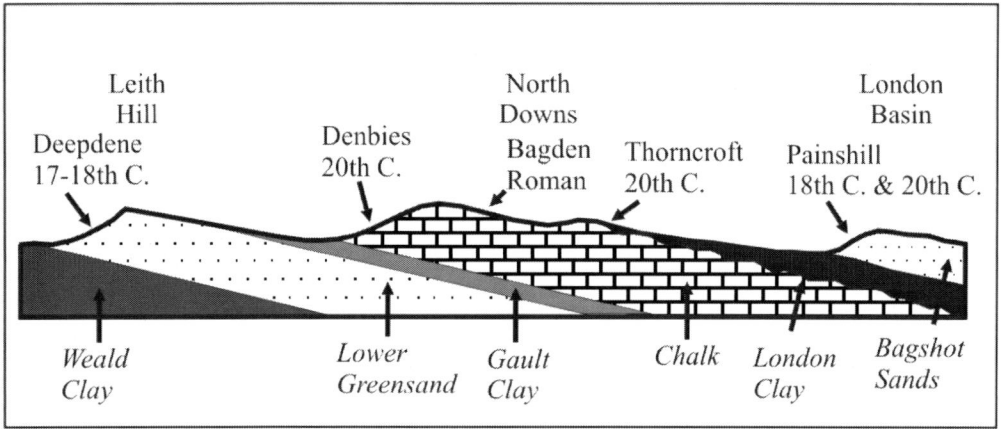

Figure 5.4 *South – north cross-section from the Weald and into the London Basin to show how vineyards through the ages, though located on various rock formations, have been planted on southerly slopes to optimise the amount of sunshine that the vines receive (From Selley, 2002).*

Water at the foot of a slope is particularly advantageous for vineyards for several reasons. First, they may reflect sunshine up the slope, thus increasing solar radiation. This effect has been noted in the Mosel and Rhine river valleys of Germany and the Neusiedler See in Austria (Jackson, 2000). Painshill vineyard, Cobham, described earlier, and Sharpham on the River Dart (Figures 5.5 & 5.6), benefit from this situation, both having water at the foot of their southerly slopes. Secondly, water gives rise to the so-called 'lake effect' described from the Great Lakes of southern Canada. Here in the autumn westerly winds crossing the lakes warm the air circulating over the ripening harvest. In winter these same moist winds bring snowfalls over the vineyards, not only killing off insects, but also providing a protective cover. It is indeed a strange sight to see the vineyards around Niagara blanketed in snow. In spring these same cool winds delay budburst from damaging late frosts.

Thus topography, controlled by rock type and climatic process, plays an important part in providing favourable sites for viticulture. Not all British vineyards, ancient or modern, have been planted with landscape in mind. Lamberhurst, in the Weald of Kent, lies on a northerly slope. Breaky Bottom, in East Sussex, occupies a north easterly-aligned valley in the Chalk of the South Downs.

Figure 5.5 *Air photo of the Sharpham promontory on the River Dart, Devon, looking west. Sharpham vineyard was cunningly planted to take advantage of the sunny southerly slope (Left hand side of the promontory), enhanced by reflected solar radiation from the River Dart (Courtesy of the Sharpham Partnership Ltd.).*

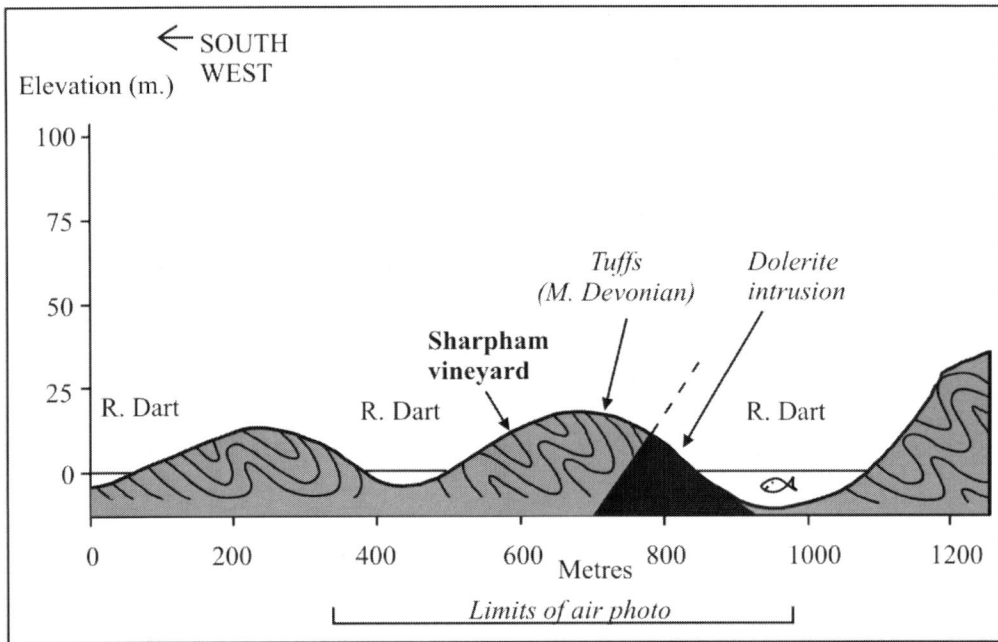

Figure 5.6 *Geological cross-section of the Sharpham promontory on the River Dart, Devon. Sharpham vineyard was planted on slates and tuffs (waterlain volcanic detritus) of Middle Devonian age. The dolerite intrusion lies beneath the copse on the right hand (North eastern) shore of the Sharpham promontory.*

On a more regional scale it is noteworthy that there are many vineyards on the Hastings Beds (Lower Cretaceous) of the High Weald, and few on the heavy clay soils of the Low Weald. The Cretaceous Chalk of southern England is particularly favourable for viticulture for reasons discussed below.

Southerly facing escarpments provide enhanced solar radiation as discussed above. Thus the North Downs and the southerly slopes of the Vale of the White Horse (Vale of Pewsey) are ideally situated. The east-west chalk ridge forming the spine of the Isle of Wight is another such terrain, of which Adgestone vineyard takes advantage. The Isle of Purbeck is also a prime area for viticulture, with southerly limestone slopes on the Chalk of the Purbeck Hills, and on the Purbeck Limestone ridge (Upper Jurassic) between Swanage and Kimmeridge Bay. In the Isle of Wight and the Isle of Purbeck though, too close a proximity to the sea is not good, as vines do not appreciate saline winds. Elsewhere inland the southerly slopes of the Chalk Downs, and the Jurassic limestones of the Cotswolds are often too high and exposed to be ideal, though local dry valleys, whose origin is discussed below, provide suitable local climatic conditions.

This brief account shows that there is no simple correlation between rock type and viticultural success. Indeed figure 5.7 illustrates how English and mainland European vineyards have been planted on rocks of many different types and age. As demonstrated above, however, rock type has a strong though indirect influence on viticulture. It controls topography, and hence microclimate. It controls the chemical and physical properties of the soil in which the vines grow. Reference has already been made to the Cretaceous Chalk limestone. This rock type is particularly suited to viticulture, and provides a good case history to demonstrate how rock type relates to topography and soil character, and thus interfaces with viticulture.

GEOLOGICAL PERIOD		ROCK FORMATION		VINEYARDS	
				BRITISH	MAINLAND
RECENT		Glacial, periglacial & alluvial superficial deposits		Morville Hall Thorncroft	
TERTIARY		Clays & sands		St Andrews Painshill Hale	Bordeaux
CRETACEOUS		Chalk		Adgestone Godstone Denbies Breaky Bottom Hambledon	Champagne
		Upper Greensand Gault Clay Lower Greensand Weald Clay Hastings Sands		Deepdene Purbeck Lamberthurst	
JURASSIC		Purbeck & Portland L-stns Kimmeridge Clay			
		M Jurassic L-stns			Burgundy
		Lias Clays			
TRIASSIC	NEW RED S-STN	Mercia mudstone Sherwood S-stn.			
PERMIAN		Sands Dolomites & shales			
CARBONIFEROUS		Coal Measures		Wootton Pilton Eglantine	
		Main L-stn.			
DEVONIAN	OLD RED S-STN	S-stns & Shales		Three Choirs Sharpham Beenleigh Manor	Beaujolais Rhine & Moselle
SILURIAN ORDOVIAN CAMBRIAN PRECAMBRIAN		Assorted horrible igneous and metamorphic rocks		La Mare Grace Dieu	Muscadet Port

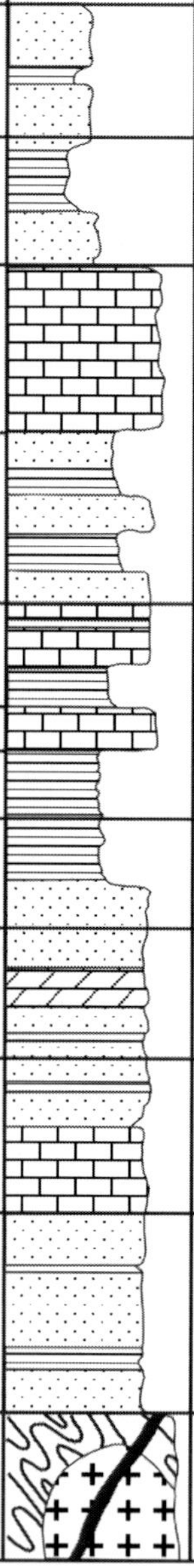

Figure 5.7 Diagrammatic rock sequence of Britain and Euroland, oldest rocks at the base, youngest at the top (As they normally occur in nature, unless complexly folded). The thicknesses of the units indicate neither the relative thickness of the strata, nor the relative duration of time. This figure also shows how vineyards grow on almost every type of rock. Superficially this suggests that geology is unimportant to viticulture, but, as the text demonstrates, geology controls everything from vineyards to volcanoes.

5.2 CASE STUDY: THE CONTROL OF GEOLOGY ON VITICULTURE EXEMPLIFIED BY CHALK

Many limestones (rocks composed mainly of $CaCO_3$) are hard tight splintery rocks, which though extensively fractured, and thus well drained, seldom have much porosity in which to store moisture. Chalk is a limestone with particular petrophysical properties in terms of its porosity and permeability. Chalk is eponymously chalky. It is lighter and softer than most limestones. This is because it is highly porous, with up to 40% void space. These pores are microscopic in size and have the ability to store large quantities of water. Chalk, however, is also commonly extensively fractured (Fig. 5.8).

Figure 5.8 Outcrop of Chalk limestone (Upper Cretaceous). Dorking, Surrey.

Chalk is normally able to absorb large amounts of water, because of its high micro-porosity, but it is normally well-drained by an extensive fracture system.

These fractures give the rock a high permeability, allowing good drainage. Chalk slopes are thus particularly good sites for viticulture because they are always well drained, but can store large amounts of moisture that the vines may tap during dry hot summers (Hancock, & Price, 1990). Chalk differs from other limestones, and is chalky, because it is composed of the microscopic fossil plates of a particular group of calcareous 'algae' called the Coccolithophoridae (not a word to pronounce after taking refreshment) (Fig. 5.9). These are composed of the mineral calcite ($CaCO_3$), a variety of lime that is stable at the temperatures and pressures that are found underground. Most lime sediment, however, is composed of the mineral Aragonite. This has the same composition as calcite, but a different crystal structure. The aragonitic form of calcium carbonate requires less energy to form than calcite, so is favoured by most carbonate-secreting animals and plants. On burial Aragonite quickly recrystallizes to the stable form of $CaCO_3$, calcite, resulting in typical hard tight limestone. If you are worried about that last bit of science forget it. If on the other hand you want to know more look it up in a textbook of sedimentology such as Selley (2000).

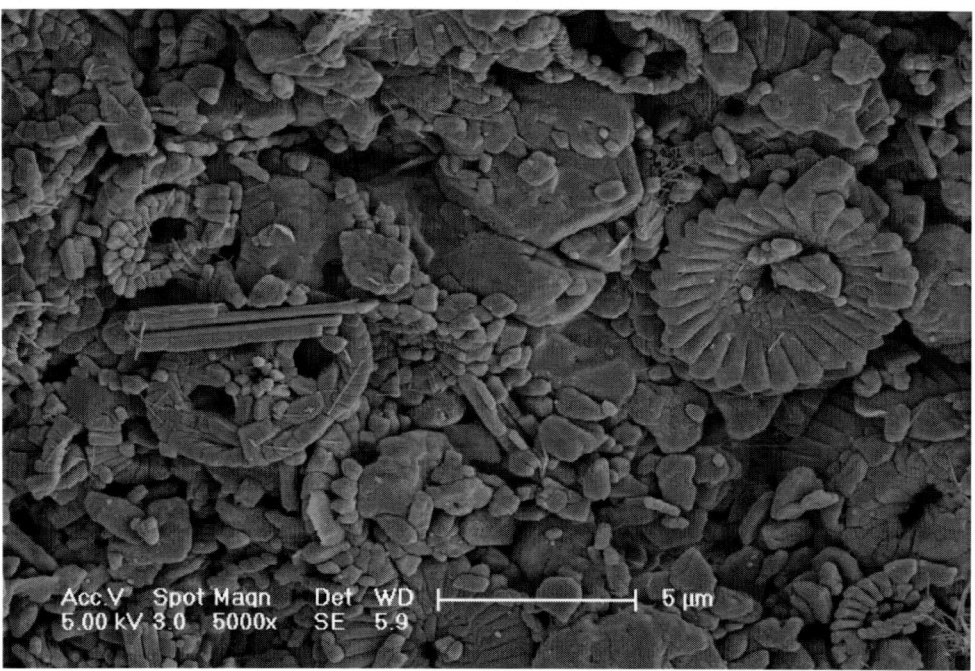

Figure 5.9 *Scanning electron micrograph of Cretaceous chalk, showing the skeletal structure of the calcitic coccolithophorid fossils of which it is composed, and the high porosity between the fossil fragments. (Courtesy of J. Huggett of Petroclays)*

Because of these special physical properties the chalk is particularly favoured for viticulture, both in France, and England. In France the Chalk, or *Craie*, crops out in two main areas, around the rim of the Paris basin, and along the north-eastern rim of the Aquitainian basin. Both areas are, confusingly, termed 'champagne'. The word was derived from the Campania region, north of Naples, which in turn took its origin from the Latin *Campus*, meaning an open field. The Chalk gives rise to open rolling countryside.

In the Charente region of southwest France the Chalk crops out in a north-west to south-east belt some 250 km long by 40 km wide on the north-east flank of the Aquitaine basin. Older strata crop out to the north-east, younger to the south-west. Brandy is produced in this area centred on the town of Cognac. The main theme of this book in general, and of this chapter in particular, is to impress on the reader the importance of geology. Here it is necessary, however, to tell of a terrible geological practical joke that was played on the vignerons of Cognac. Professor Henri Coquand (1813-1881), sometime President of the Geological Society of France, was a scientist of great distinction. He carried out extensive fieldwork in the Charente at the end of which he ran a field trip for the Geological Society of France. This trip concluded on 13 September 1857 with *'un diner splendide'* at the Chateau Malberchie. Now it was, and still is, the habit of learned scientific societies to hold splendid dinners, which conclude with a talk by one of the diners. On this evening the dinner was so splendid that the good professor thought it a suitable occasion to *'extracter le Michel'* from the brandy vignerons of the area. He gave a humorous talk advancing the ridiculous hypothesis that the quality of brandy declined in ever increasing circles centred on Cognac, and that this was due to the 'chalkyness' of the soil (Figure 5.10).

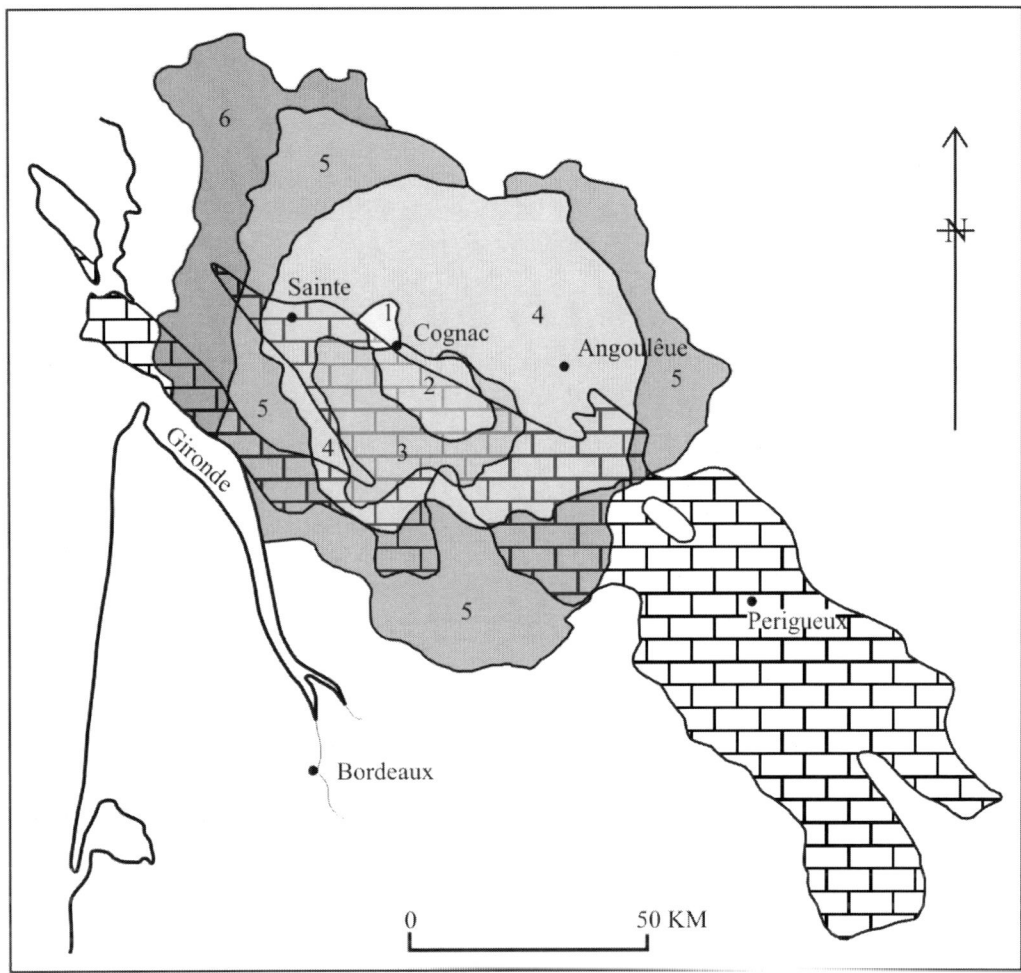

Figure 5.10 *Coquand's geological jest of 1857. Map purporting to show how the quality of brandy (decreasing from 1-6) decreases in ever increasing circles from Cognac with decreasing 'chalkiness' of the soil. This is clearly nonsense since the Chalk that gives the soil it's 'chalkiness' crops out in a belt that cross-cuts the brandy bulls-eye. Chalk (Cretaceous) shown by 'brick' ornament.*

This is demonstrably ridiculous as the strata crop out in belts cross-cutting the 'bulls-eye'. It may also be thought incredible that the chalkyness of the soil imparts a character to brandy that survives the distillation process. Professor Coquand and his colleagues were so diverted by this geological jest, that on his return from the field, he included it in his otherwise serious report of the excursion in the Bulletin of the Geological Society of France (Coquand, 1857), where, no doubt, the joke was enjoyed by a wider readership (Hancock, 1999a, Hancock and Selley, 2003).

Unfortunately, some years later, wine experts came across the paper, failed to recognise it as a joke, and have solemnly replicated it ever since. The last laugh, however, is on the geologists, who, a century later, recaptured the 'bulls-eye' from the vignerons and accepted it as gospel (Wilson, 1998).

Moving swiftly from the Charente to northern France, the Chalk crops out again around the rim of a saucer-shaped basin of sedimentary rocks centred on Paris. The north-eastern flank of this basin is the home of the only sparkling white wines that may legally be termed Champagne. In this region the chalk escarpments reach heights of some 300m. Vines are grown on the mid – lower slopes at altitudes of less than 200m. Figure 5.11 illustrates the topography of the chalk escarpments of Champagne, and contrasts them with those of England. In England the Downs seldom reach such elevations and chalk vineyards, such as those at Adgestone, Hambledon, Godstone, Oxted and Denbies, are generally less than 150m above sea level.

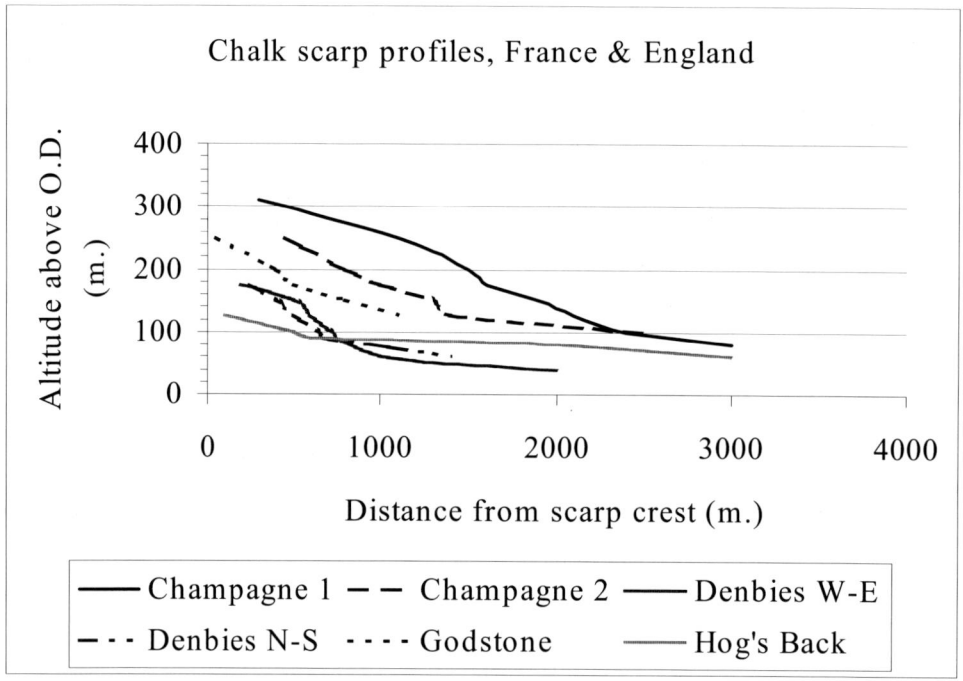

Figure 5.11 *Profiles of the chalk escarpments of Champagne, France, and of England. Note that the English escarpments, and thus the vineyards on their slopes, are lower than those of France.*

The distinctive landscape of the Chalk has been commented on by many natural scientists. The Reverend Gilbert White, author of 'The Natural History and Antiquities of Selborne' (1789), commented:

'I think that there is somewhat peculiarly sweet and amusing in the shapely figured aspect of the chalk hills.....I never contemplate these mountains without thinking I perceive somewhat analogous to growth in their gentle swellings and smooth fungus-like protuberances, their fluted sides, and regular hollows and slopes, that carry at once the air of vegetative dilation and expansion. Or was there ever time when these immense masses of calcareous matter were thrown into fermentation by some adventitious moisture..........'

Rev. G. White. Letter to the Hon. Daines Barrington. 9 December 1773

Though one of the greatest naturalists of his day, Gilbert White's account displays greater powers of observation and expression than of deduction.

A distinctive feature of the Chalk landscapes of England and France is the occurrence of so-called 'dry valleys'. These are valleys that, as their name implies, lack a river. This poses the question of how can a valley be eroded in the absence of a river? The answer lies in the past climate of the landscape. During the Ice Age of the last million years glaciers expanded and contracted several times across Scandinavia and northern Britain, but did not extend to the south of the Thames Valley. Southern England was tundra, with frozen ground, termed permafrost. During these periglacial climatic conditions the ground was frozen to a depth of several metres. In summer, however, the surface of the chalk would warm sufficiently for the ice contained within fractures and smaller pores to thaw. Unable to seep down through the permafrost, the melt water would flow down hill carving river valleys on its way to the sea. At the end of the last Ice Age the permafrost melted. Surface water is now free to sink down through the fractured chalk leaving the valleys dry (Fig.5.12). Dry valleys are important to viticulture. They provide sheltered locations within which vineyards may be planted, irrespective of the regional topography. Thus the South Downs of Sussex have steep northerly-facing shady escarpments, and an exposed gentle southerly dip slope. Nonetheless vineyards such as Breaky Bottom, near Lewes, find shelter in dry valleys.

Figure 5.12 Typical chalk topography near Dorking, Surrey, looking eastwards towards Box Hill, beyond which extends the sun-kissed southerly-facing scarp of the North Downs. At the foot of the escarpment is the low ground of the Weald. The gap in the North Downs has been cut by the River Mole, so-named because of its habit of disappearing down fractures in the chalk in times of drought. In the foreground is the easterly trending periglacial dry valley of the Dell, the most sheltered part of Denbies vineyard.

Though limey soils are good for viticulture there is one particularly problem to which limestone soils in general, and the chalk in particular, are prone. This is a yellowing of the leaves due to a lack of chlorophyll. This condition, termed chlorosis, is caused by excessive alkalinity, and is described in more detail in the next section.

The Chalk limestone of the Downs of southern England is extensively overlain by a thin veneer of clay with flints, formed by poorly understood processes during the Ice Age. At the foot of the Downs alluvial gravels, sands and clays of a river valley often overlie the Chalk. Weathering produces a layer of soil that overlies both solid bedrock and superficial deposits. Weathering is the term applied to the break down of rock by physical, chemical and biological processes. Weathering may result from physical processes, such as the fracturing of rocks by freeze-thaw action in glacial climates, and the alternating thermal expansion and contraction of rock in hot deserts. Weathering may also result from chemical processes, such as the dissolution of minerals by acid rain. (Contrary to some modern views there is nothing new in acid rain. Rainwater has always been acidic, as demonstrated by ancient cave systems dissolved out of limestone.) Weathering may also result from biological processes such as the fracturing of rocks by plant roots, and the digestion of tasty minerals by bacteria.

The rate and type of weathering is closely related to climate. The interplay of climatically controlled weathering, and the underlying superficial deposits or bedrock determines the character of the overlying soil.

5.3 THE ANSWER LIES IN THE SOIL

The soil on which vines grow is of great importance. Indeed soil has been poetically described as 'the soul of the vine'. Soil results from the interplay of climate and rock type. Soil may be defined as the disaggregated detritus formed by the weathering of bedrock and/or superficial deposits. It is the insoluble residue of the underlying parent material. Soil contains organic matter of various types, ranging from rotted tree roots to bacteria, which together constitute humus. Soil contains organic matter in varying amounts, ranging from the waterlogged peat soils of cool temperate climates, to the inorganic soils of deserts. There is general agreement on the physical properties of soil that are beneficial to viticulture, but less agreement on the chemical composition of the soil that is favourable. Vines like a steady, but moderate, supply of moisture. They do not thrive in waterlogged soils. This is one reason why vineyards are often planted on hillsides since they allow good drainage. In dry conditions vines develop taproots that may penetrate down through fractures for many metres to reach the water table.

One last physical attribute of soil to note is colour. Soils come in a wide range of colours that depend on the bedrock and climate. Dark soils reflect less sunlight than pale ones, but absorb more heat and thus warm up during the daytime and can radiate heat back into cold night air to the benefit of ripening grapes. In many German vineyards, which are planted near the northern climatic limit of viticulture, the indigenous black slate around the roots is favourable, and sometimes black rocks, such as basalt, are scattered around the roots of the vines.

Turning from the physical properties of soil that favour viticulture to the chemical composition, there is less consensus. The only parameter on which all agree is that high nitrogen content in the soil is bad, thus manuring is not a good idea. Soils vary from acid to alkaline (for non-chemists: acididity/alkalinity is measured in terms of pH from $0 - 14$, where 7 is neutral, anything less is acidic, anything more is alkaline).

Vines do not thrive in extremely acid soils with a pH less than 5. Such soils are usually to be found in regions of high rainfall. Alkaline soils, with a pH above 8.5, usually have free lime or salt. The latter is obviously to be avoided. Alkaline soils occur on limestone, naturally. Though some lime is good for plant growth, excessive lime can inhibit the uptake of nutrients essential for plant growth. This may cause the leaves to turn yellow due to a lack of chlorophyll, a condition known as chlorosis. Chlorosis is due to deficiencies of several important trace elements, but especially iron. It can be cured with the addition of iron. Iron occurs naturally in many rocks, in pyrites (Iron sulphide), in chalk, or in glauconite (a complex potassium ferro-magnesian silicate) found in greensand, and some chalk. Where iron is absent in the bedrock it may be applied in the form of chelated iron. Chlorosis may be avoided by planting lime-tolerant vines. The optimum amount of potassium for viticulture varies with climate and rainfall. Figure 5.13 shows how the interplay of climate and geology controls soil: 'the soul of the vine'.

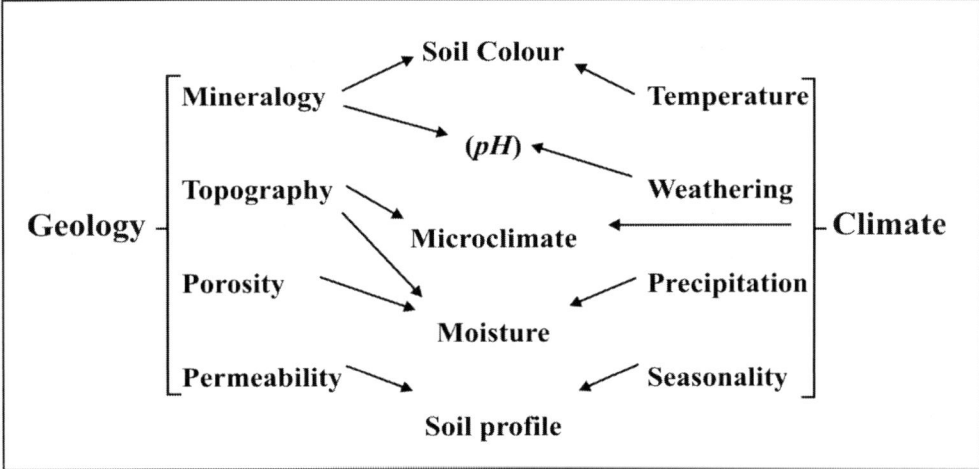

Figure 5.13 *Soil, the "soul of the vine" (Wilson, 1998), results from the complex interplay of geology and climate. Note that in many vineyards it is the superficial deposits, such as alluvium, wind blown sand or glacial boulder clay, that determine the character of the soil, as much as the solid bed rock beneath.*

Hyams (1949) helpfully produced a Soil Comparison Table that listed the soils of the French winegrowing areas, and offered the nearest English equivalent. Interestingly his detailed catalogue of the French wine-growing areas omitted Champagne and the Chalk. The sparkling white wines of southern England are now international prize winners, and the Cretaceous chalk Downs highly prized for viticulture

Of the granite soils of Burgundy, on which the Gamay grape is grown, Hyams noted that the moors of Devon and Cornwall are the nearest English equivalent, and commented 'Who will investigate the possibility of turning these wastelands into vineyards?' Dartmoor Vineyard was planted in 1992, albeit on the sedimentary rocks of Bovey Tracey on the eastern side of the moor. As discussed in Chapter 7, if global warming continues, it may not be many years before someone plants a vineyard on the Dartmoor granite itself.

5.4 THE WAR AGAINST 'TERROIR'

The French have subsumed the concatenation of soil, topography and climate into a quasi-mystical concept that they term 'terroir'. 'Terroir' is a key parameter in the Appellation Contrôlée scheme of classifying French wines. The relationship between 'terroir' and wine has been expressed in a simple mathematical formula by Jefford (2003) as:

$$Wine = grape\ variety + terroir + vintage + winemaking$$

Jefford (*ibid.*) writes poetically that 'Fermentation seems to draw out the grape's memory of the stone in which it once buried its roots, of the view across the valley that it had as it ripened, and the ever-changing weather pattern it enjoyed or endured as the season unfolded.' Geologists will concur, though mentally replace 'stone', which is something worn on a ring around the finger, with 'rock'. 'Terroir' has attracted as much argument between *Terroiristes* and *Anti-terroiristes* as did the medieval theological debates on the number of angels who could pirouette on a pinhead. It has been used in the same way as a drunk uses a lamppost, more for support than illumination, and as a method of forcing up the price of vineyards (Moran, 1998).

Hancock (1999a) and Huggett (2006) analysed the 'terroir' concept from the geological perspective. They point out that vine root systems may penetrate into rock for several metres, but these deep roots are primarily seeking water. Vines derive most of their nutrients from depths of less than 0.6m. In other words it is the chemistry of the soil that may be important in imparting flavour to wine, not the bed rock. In some winelands the soil is produced by the *in situ* weathering of bed rock. In many winelands, however, the bed rock is overlain by superficial deposits of gravel, sand or clay of glacial, eolian or fluvial origin. It is therefore, the chemical and physical properties the superficial deposits that may affect the character of vine and wine, not the character of the bed rock beneath. In summary geology has much to do with viticulture, but little to do with vinification. The control of geology is strong, but indirect. The interplay of geology and climate determines the soil in which a vineyard grows, and the landscape in which it stands.

No Anglo-Saxon, however, could possibly understand such a quintessentially French concept as 'terroir', and it would be presumptuous to attempt to do so, as Hancock remarked (1999a) the concept of 'terroir' combines medieval mysticism with second-rate science. Nonetheless geologists will continue to research geoviticulture as it provides one of the few careers in which you can drink on the job.

5. 5 BIBLIOGRAPHY

There are several excellent books to introduce the general reader to geology. For example:

Fortey, R. 2004. The Earth – An intimate History. Harper Collins. London. 480pp.

Redfern, M, 2000. Origins. The evolution of continents, oceans and life. Cassell & Co. London 360pp.

Redfern, M. 2003 The Earth: A Very Short Introduction. Oxford University Press. Oxford.141pp.

5.6 REFERENCES

Bargman, CJ 2005. Geology and wine in South Africa. Geoscientist. 15. 4-6.

Bezak, V., Suk, M. and B. Molak. 2002 Rocks and wines in Slovakia. European Geologist. 13. 35 – 38

Coquand, H. 1857. Réunion extraordinaire à Angoulême du 6 au 14 Septembre 1857. Bull. de la Société Géologique de la France. 14. 841 – 903.

Hancock, J. M. 1999. What makes good wine? Science Spectra. 15. 74 - 79.

Hancock, J. M. 1999a. Terroir, The Role of Geology, Climate and Culture in the Making of French Wines. Jl. of Wine Research. 10. 43-49.

Hancock, J.H & Price, M. 1990. Real Chalk balances the water supply. Jl. Wine Research. 1. 45-60.

Hancock, JH & Huggett, JM. 2004. Geological Controls in Coonawarra. Jl. Wine Research. 15. 112-122

Hancock, J.H. & Selley, R.C. 2003. Coquand's joke. Geoscientist. London. 13. 10. 17

Huggett, JM. 2006. Geology and wine: a review. Proc. Geol. Ass. Lond. 117. 239-247

Hyam, E., 1949. The Grape Vine in England. The Bodley Head. London. 208pp.

Jackson, R.S. 2000. Wine Science: Principles, Practice, Perception. Academic Press. San Diego. 648pp.

Jefford, A. 2003. Rooted to the Spot. Waitrose Food Illustrated. May 2003. 72 – 74.

Macqueen, RW & Meinert, LD., 2006. Fine Wine and Terroir   The Geoscience Perspective. Geoscience Canada Reprint Series 9. Geological Association of Canada. St John's. 247pp.

Maltman, A. 2003. Wine, beer and whisky: the role of geology. Geol. Today. 19. 22-29

Moran, W. 1988. The wine appellation: environmental description or economic device? 2nd Internat. Cool Climate Viticulture & Oenology Symp. Aukland NZ. 356-360

Pomerol, C. (Ed.) 1989. Terroirs & Vins de France. 2nd Edn. (English translation published by R. McCarta)

Selley, R.C. 1996. The Geological Inspiration for the Denbies Vineyard. Denbies Wine Estate. Dorking. 8pp

Selley, R.C. 2000. Applied Sedimentology. (Second Edn.) Academic Press. San Diego. 521pp.

Selley, R.C. 2002. Geological and climatic controls on English wines. Geol. Assoc. Mag. London. 1.2. 6-8.

Smart, R.E. 1999. Hillside vineyards. In: The Oxford Companion to Wine. J. Robinson (Ed.) Oxford University Press. Oxford. p.351

Wallace. P. 1972. The Geology of Wine. Proc. 24th Internat. Geol. Cong. Montreal. Section 6. 359-365.

White, Rev. G. 1789. The Natural History and Antiquities of Selborne. Bickers & Sons. London. 568pp. (and reprinted in various formats and by various publishers ever since)

Wilson, J.E.1998. Terroir. Mitchell Beazley. London. 336pp.

CHAPTER 6.

Viticulture in the Industrial Revolution Warm Phase

It is a truth universally acknowledged, that an estate in possession of a southerly prospect, must be in want of a vineyard.

JAMES AUSTEN (1813)

6.1 THE RENAISSANCE BEGINS

The penultimate chapter showed how, during the Little Ice Age of the 15th to 19th centuries, viticulture continued in southern England as a hobby for eccentric gentlemen. In the 20th century the industrial revolution accelerated, the climate began to warm, and the renaissance of viticulture began. These three facts may, or may not, be interconnected.

The renaissance of British viticulture began, not in southern England, as might be expected, but in south Wales. In 1875 the incredibly wealthy Lord Bute established a vineyard in the grounds of Castle Coch, outside Cardiff. The site was ideal, with a sheltered well-drained southerly slope. The soil was deeply weathered New Red Sandstone unconformably overlying Carboniferous Limestone at a depth of less than a metre. Some 2,000 vines were imported from France and planted over the 3-acre site. The noble lord's plans were greeted with derision. The comic magazine 'Punch', the 'Private Eye' of its age, stated that it would take 4 people to drink it: the victim, two men to hold him, and a fourth to pour it down his throat. By 1881, however, white wine, described as 'still champagne' (sic) was selling at £3 a dozen. By the late 1880'ies production was up to 3,600 bottles per annum, and Lord Bute planted over 5 acres of additional vines. Red wine was now added to the range of wines offered. By 1905 63,000 vines were flourishing.

Tragically it was not to last. Wine making ceased on the outbreak of the First World War, and the vines were uprooted in 1920. Despite the success of Welsh wine, two world wars and the intervening years of the depression inhibited the reestablishment of viticulture, though it might have been just the thing to cheer everyone up. The next quantum leap was the establishment of a vineyard at Oxted, Surrey, by Ray Barrington-Brock. Barrington-Brock was a scientist by training and employment, but a gardener by bent. After spending some time researching viticulture on the European mainland, he established the grandly named Oxted Viticultural Research Station on a south-facing slope of chalk at a height of 450 feet (135m).

This project was, as its name suggests, a genuine research station, not just a few vines planted by an enthusiast of the 'muck and mystery' school of horticulture. Barrington-Brock experimented with many different German and French grape varieties, soils, fertilizers, and styles of pruning and training. Between 1949 and 1964 the results were published in a series of research reports.

In 1951, stimulated by Barrington-Brock's reports, Sir Guy Salisbury Jones established a commercial vineyard at Hambledon, Hampshire. It was 30 years since the Castle Coch vineyard was grubbed up in 1920. 4,000 French vines were planted on one and a half acres of exposed south-facing chalk slope at an elevation of some 300ft (100m). The first wine was produced in 1955. Hambledon vineyard attracted much national attention and other heroes began to plant vineyards across southern England. Over the years English winemakers changed from atavistic eccentrics to agricultural entrepreneurs.

Though Britain is currently at the northern limit of wine production, three factors are in its favour. First, the evolution and careful selection of cultivars that can withstand the climate. Many of these are German or French hybrids that are mainly suitable for white wine production. Good red wine has been very difficult to produce, but this is now changing with the introduction of Rondo, a variety that produces excellent red wine in marginal climatic conditions. Improbably Rondo originated in Manchuria and has migrated westwards from Asia to Europe along the northern climatic limit of viticulture. The second factor that has favoured viticulture in Britain has been the application of ways of overcoming the effects of late frosts that can damage vines once the buds have already burst. These methods include wind turbines, though these produce a noise that disturbs the natives, heaters fuelled by diesel oil (Denbies) (Figure 6.1), sprayed on rubber latex, and even polytunnels (Worthenbury, Offa's, Beenleigh Manor) (Fig. 6.2). The third factor that may favour British wine production is climate change. If global cooling takes place we are all doomed. But if global warming occurs then the immediate prospects for British (English, Welsh and Scottish) viticulture are excellent. This aspect is examined at some length in the next chapter.

Figure 6.1 *Photo showing some of the 600 diesel oil heaters used to prevent budding vines from damage by late spring frosts on the river terrace of the River Mole. Denbies vineyard, Surrey. Once it used to take all night to light the heaters. They are now largely redundant as biodegradable latex is sprayed on the vines instead.*

Figure 6.2 *Method of overcoming late frosts. Merlot and Cabernet Sauvignon vines growing under polytunnels at Beenleigh Manor vineyard, Devon.*

These three factors led to the planting of ever more vineyards through the latter half of the previous century. The increasing number of commercial wine producers led to the formation of the English Vineyards Association in 1967 with about a dozen members. Ten years later, by 1977, there were 124 vineyards of >1 acre (0.4 Hectares) in size in England, Wales, Ireland and the Channel Islands (Barty-King 1977). In 1996 the English Vineyards Association metamorphosed into the United Kingdom Vineyards Association. Now (2004) there are some 400 vineyards covering some 850 hectares, and 50 wineries in the United Kingdom, including England, Wales, the Channel Islands, Northern Ireland, but not Scotland – yet (Figure 6.3 and Table 6.1) (Skelton, 2001, Mann, 2002, Johnson, 2004).

The most northern vineyard at the time of writing (2003) is Mount Pleasant in Yorkshire, at Latitude 53^0 45'. This is about the same latitude as Edmonton, Canada, though it is not remarkable bearing in mind that there are now vineyards in Norway and Finland. Long may the Gulf Stream flow. The area under cultivation varies from year to year, and according to whether the figures includes established vineyards, or recently planted ones that have yet to yield their maiden vintage. Similarly production varies from year to year, largely according to the weather, but is now in the region of 3,000,000 bottles of wine a year. For the latest information on the status of British viticulture visit www.englishwineproducers.com.

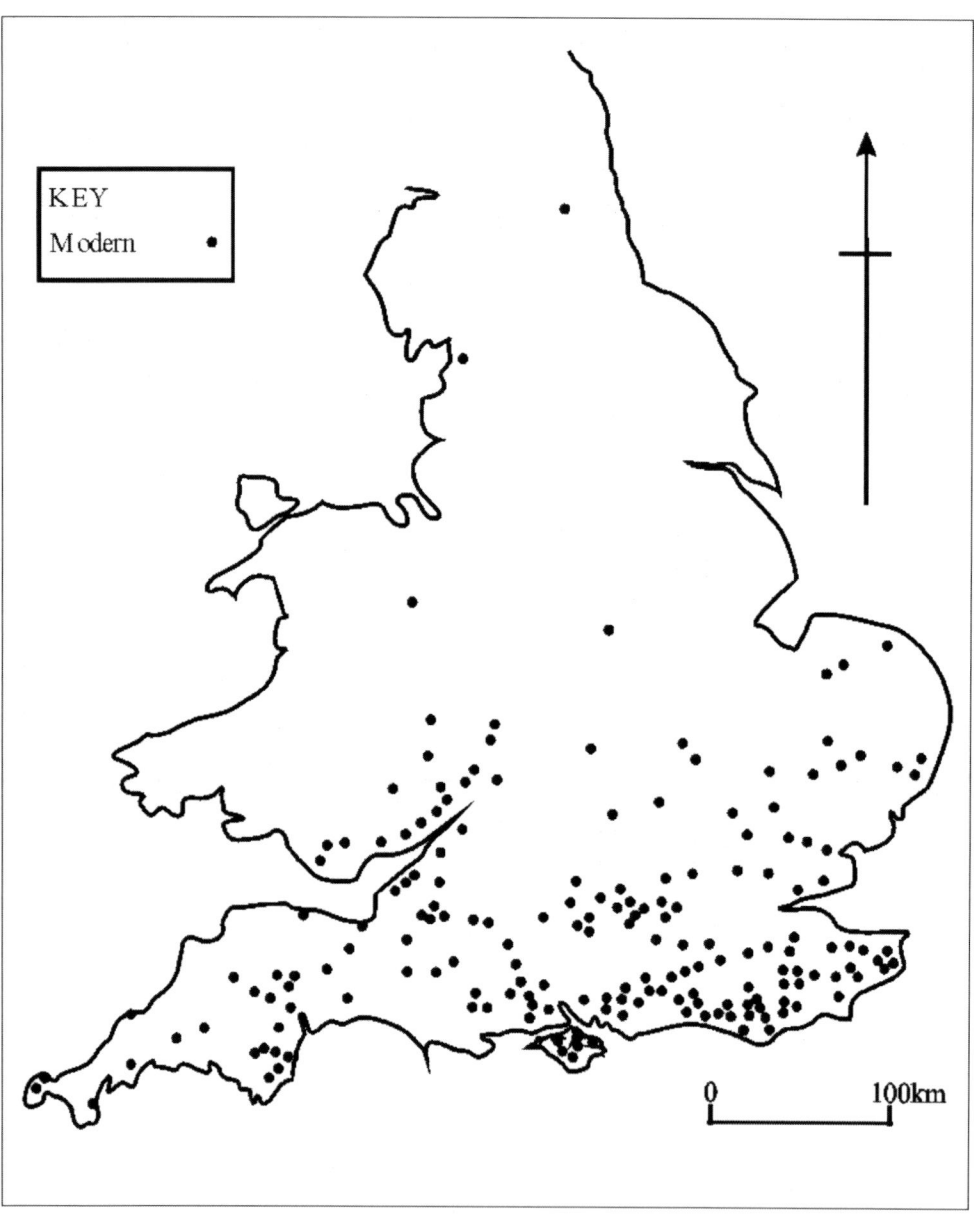

Figure 6.3 *The distribution of British vineyards in the present day Industrial Revolution Warm Phase. Vineyards have now advanced from the enclave of south-east England, to which they had retreated in the 'Little Ice Age', to reach the Humber – Severn boundary of Roman and Medieval viticulture and beyond. Since the publication of the first edition of this book in 2004 the most northerly vineyard has advanced from Mount Pleasant in Lancashire, to Acomb, Yorks., within a few miles of Hadrian's Wall. (Compiled from several sources, but principally Skelton, 2001, Mann, 2002, Johnson, 2004).*

COUNTY	NAME	SIZE	COUNTY	NAME	SIZE
Berks.	Cane End	5.53	Essex	Bardfield	1
	Domaine Madeleine	0.65		Carter's	1.71
	Lillibrooke Manor	0.11		Castle	0.2
	North Court Farm	0.1		Colne Valley	0.58
	Valley	8.24		Felsted	4.25
	Windsor Forest	1		Great Stocks	2.63
Bristol	Avonwood	0.25		Little Witney Green	0.45
	Conham Vale	0.06		Mersea	2.67
Bucks.	Appledown	0.007		Mower	0.08
	Bower Farm	0.07		Nevards	0.35
	Claydon Assn.	0.0297		New Hall	36.3
	Hale Valley	0.62		Olivers Farm	0.02
	Milton Keynes	2		Potash	0.1
	Warden Abbey	2.26		Stony Hills	0.99
Cambridge	Chilford Hall	7.42		Writtle College	0.28
	Elysian Fields	0.06	Gloucs.	Compton Green	1.5
	Meldreth	0.23		Kents Green	0.18
	Shelford Green	0.07		Little Foxes	0.28
Channel Islands				Saint Anne's	2.11
Jersey	Chateau le Catillon	3.67		Three Choirs	30
	La Mare	6.9		St Augustine's	0.4
			Hants.	Beaulieu	0.62
Cornwall	Bosue	0.41		Bishops Waltham	1
	Camel Valley	1.8		Braishfield Manor	2.47
	Durra Valley	0.04		Buddlemead	0.08
	Lambourne	0.89		Coach House	0.33
	Penberth Valley	0.45		Court Lane	0.51
	Polmassick	0.5		Danebury	2.15
	Porthallow	0.68		Hale Valley	0.25
Devon	Beenleigh Manor	0.13		Hamble Valley	0.29
	Dartmoor	0.2		Jays Farm	1.7
	Down St Mary	1.52		Marlings Valley	0.72
	Follymoor	0.56		Meon Valley	6.48
	Manstree	1.41		Northbrook Springs	5.1
	Oakford	1.05		Priors Dean	0.36
	Rock Moors	0.24		Setley Ridge	1.68
	Sharpham	3.96		Titchfield	0.89
	Summer Moor	1.73		Waterwynch	0.04
	Weir Quay	0.04		Webb's Land	3
	Yearlstone	3.04		Westward House	0.1
Dorset	Horton Estate	3.63		Wickham	7.28
	Lytchett Matravers	0.03		Wooldings	4.93
	Parhams	0.4	Hereford.	Beeches	0.12
	Purbeck	0.29		Bodenham	3.86
	Wake Court	1.27		Coddington	0.96

Table 6.1 *British vineyards of the Industrial Revolution Warm Phase (Largely from data in Skelton, 2001, by courtesy of the author. See also Mann 2002 and Johnson, 2004.*

COUNTY	NAME	SIZE	COUNTY	NAME	SIZE
Hereford (Contd.)	Frome Valley	1	Kent (Contd.)	Swattenden Ridge Farm	0.25
	Sunnybank Nursery	0.34		Tenterden Vineyard	7.57
	Treago	0.3		Throwley	1.38
	Wyecliff	0.66		Wootton Park	0.81
Herts.	Broxbournebury	1.21	Lancashire	Mount Pleasant	0.08
	Frithsden	0.81	Leicestershire	Chevelswarde	0.238
	Harpenden Garden	0.001		Grace Dieu	0.22
	Hazel End	0.83		St Peter's	0.25
	Mimram Valley	0.48		Welland Valley	0.35
	Tudor Grange	0.024	Lincolnshire	Bishop's Palace	0.09
Isle of Wight	Adgestone	3.2	London	Mill Hill Village	0.06
	Chapel Farm	8		Saint Andrews	0.194
	Rosemary	11.33	Norfolk	Harling	2.63
	Rossiters	2.43	Northants.	Ladywell	0.83
Kent	Ash Coombe	1.4		Valhalla	0.024
	Barnsole	1.39		Vernon Lodge	0.09
	Bayford House	0.07		Windmill	0.26
	Bearsted	1.6	Notts.	Eglantine	1.3
	Biddenden	8.1	Oxford	Bothy, Abingdon	1.04
	Challenden	0.62		Boze Down	2.2
	Chapel Down	N.A.		Brightwell	4.27
	Chiddingstone	14.43		Chiltern Valley	0.41
	Clay Hill	2.63		Fawley	0.34
	Conghurst	0.2		Grange Farm	0.025
	Elham Valley	0.72		Hendred	2
	Groombridge Place	1	Rutland	Dragons Rock	0.2
	Hadlow Down	0.56	Scilly Isles	Saint Martins	0.42
	Harden	5	Shropshire	Morville Hall	0.07
	Horsmonden	2.03		Wainbridge	
	Kempes Hall	0.1		Wroxeter	
	Knowle Hill	0.05	Somerset	Avalon	0.91
	Lamberhurst	10.64		Bagborough	2.43
	Leeds Castle	1.31		Bittescombe Mill	0.036
	Lewins	0.13		Castle Cary	1.7
	Mayshaves	0.16		Cheddar Valley	0.52
	Meopham	1.93		Cufic Vines	0.12
	Mount Ephraim	0.43		Dunkery	2.89
	National Vine Collection	0.24		Moorlynch	1.5
	Penshurst	5		Mumfords	1.24
	Rowenden	0.15		Oatley	1.9
	Sandhurst	5.5		Quantock	0.1
	Scott's Hall	0.2		Rodney Stoke	1.43
	Staple St James	1.3		Staplecombe	0.73
	Surrenden	1.2		Whatley	0.68
				Wootton	0.58

COUNTY	NAME	SIZE	COUNTY	NAME	SIZE
Stafford.	Halfpenny Green	6.2	Sussex E. (Contd.)	Sedlescombe organic	3.24
Suffolk	Boyton	0.81		Spilstead	4
	Bruisyard	4.05		Spring Barn	1.09
	Cavendish Manor	0.36	Sussex W.	Bookers	2.43
	Gifford's Hall	3.62		Lydhurst	0.27
	Helions	0.23		Nutbourne	6.06
	Hintlesham	0.09		Nyetimber	15.8
	Ickworth	0.8		Rock Lodge	2.83
	Melton Lodge	0.67		Standen	1.04
	Shawsgate	5.05		West Stoke	2.4
	Staverton	0.52	Wiltshire	Chalkhill	1.01
	Willow Grange	0.4		Elms Cross	2
	Wyken	2.12		Little Ashley	0.18
Surrey	Denbies	106		Sherston	0.5
	Furnace Brook	1.08		Swiss Cottage	0.019
	Godstone	1.5		Wylye Valley	3.53
	Greyfriars	0.46	Worcestershire	Astley	2
	Home Farmhouse	0.157		Parkfield	0.12
	Painshill Park	0.73		Tiltridge	0.51
	Send	1.89		Treetops	0.03
	Thorncroft	3.44	Yorkshire	Leventhorpe	2.21
Sussex (East)	Ashburnham	4		Water Hall	0.06
	Barkham Manor	14		Acomb Grange	
	Barnsgate Manor	3.55			
	Battle Wine Estate	13.36		**WALES**	
	Bewl Water	6.5	**COUNTY**	**NAME**	
	Bodiam	1.82	Monmouthshire	Brecon Court	2
	Breaky Bottom	2.23		Monnow Valley	1.28
	Brickyard	1.78		Offa's	0.06
	Burwash Weald	1.2		Sugar Loaf	1.59
	Carr Taylor	7.4		Tintern Parva	0.8
	Chingley Oast	0.69	Pembrokeshire	Cwm Deri	0.74
	Davenport	1.67	Ceredigion	Ffynnon Las	0.2
	Ditchling	2	Glamorgan	Glyndower	1.27
	English Wine Centre	0.01		Llanerch	2.2
	Hackwood	1.64	Gwynedd	Eryri	0.37
	Hidden Spring	3.04	Caerdydd	Pant Teg	0.71
	High Weald	1.51	Wrexham	Worthenbury	0.25
	Mill Oast	1			
	Plumpton College	0.46			
	Ridge View	6.48		**SCOTLAND**	
	St George's	4.49		Pending	

Table 6.1 *British vineyards of the Industrial Revolution Warm Phase – continued & concluded.*

The renaissance of English viticulture described above has been hindered by several factors, environmental, politico-economic and social. Climate, an obvious one, albeit of declining importance, is discussed in depth in the next chapter. Taxation and pests are others. A helpful government charges English wine producers the same duty per bottle as importers; currently (2004) £1.16 for still, and £1.65 for sparkling wine. It is the economies of scale that make life hard for British producers. The cost of duty is about the same as the cost of actually producing the grape juice. Many New World vineyards operate on a far larger scale and with (in some countries) much lower labour costs than in the UK. Despite transportation charges they can still compete on the supermarket shelf with aboriginal wines. All is not lost, however. Ten years ago a British wine drinker might buy a bottle of English wine out of curiosity for £6.00-7.00. The combination of taste and cost might ensure that they never repeated the experience. Now, however, prices have come down for every day drinking wines, though the costs of dessert and sparkling wines are higher, justifiably so. The quality of English wine has improved dramatically, for reasons discussed below.

The third inhibiting factor for British viticulture is pests. These range from deer to *Fomes igniarius*, colloquially known as 'Black measles', truly (Unwin, 1991). One of the downsides of global warming is the poleward advance of pests along with viticulture (Tate, 2001). Perhaps one of the biggest pests that British viticulturalists may face, however, is their local council. The case of Wroxeter Roman vineyard is one of the most celebrated/notorious examples. When the heroic Millingtons opened their vineyard Shrewsbury local authority took the view that wine making was an industrial process that required planning consent (Consider the industrial pollution and environmental degradation of the Loire Valley and the Champagne of France, for instance). It took 7 years and several public enquiries and court hearings before 3 Law Lords in the High Court took the view that to consider viticulture anything other than agriculture was 'an affront to common sense'. Denbies, described below as a case history of a modern vineyard, encountered similar problems with the local Mole Valley District Council.

6.2 CASE HISTORY: DENBIES VINEYARD, SURREY

Case histories of vineyards were described in previous chapters for each of the climatic phases of the last two millennia. Thus it is now appropriate to describe a vineyard of the current 'Industrial Revolution Warm Phase'. Denbies, on the North Downs at Dorking, in Surrey is typical of many southern English vineyards in its geological and scenic setting, range of cultivars and resultant wines.

Adrian White purchased Denbies Estate from Henry Cubitt, the Fourth Lord Ashcombe, in 1984. Included within the estate was Bradley Farm, noted for growing pigs, maize and blackberries. For some years an eccentric local geologist, interested in wine as geologists often are, had stared out from his study window at the slopes of Denbies Down. Living on the backside of Charles Howard's Deepdene vineyard, he was aware of the history of viticulture in the Surrey hills. This geologist also knew that the North Downs and the Champagne area of France shared a common rock type, the Cretaceous Chalk limestone. He had a geospasm and wrote an unsolicited report for Adrian White, pointing out the tradition of viticulture in the area, the fact that Bradley Farm possessed geology, and hence soil conditions, similar to the French Champagne (Fig. 6.4). Furthermore the topography was ideally suited too, with well-drained southerly and south-easterly slopes protected from the prevailing westerly winds by a projecting headland.

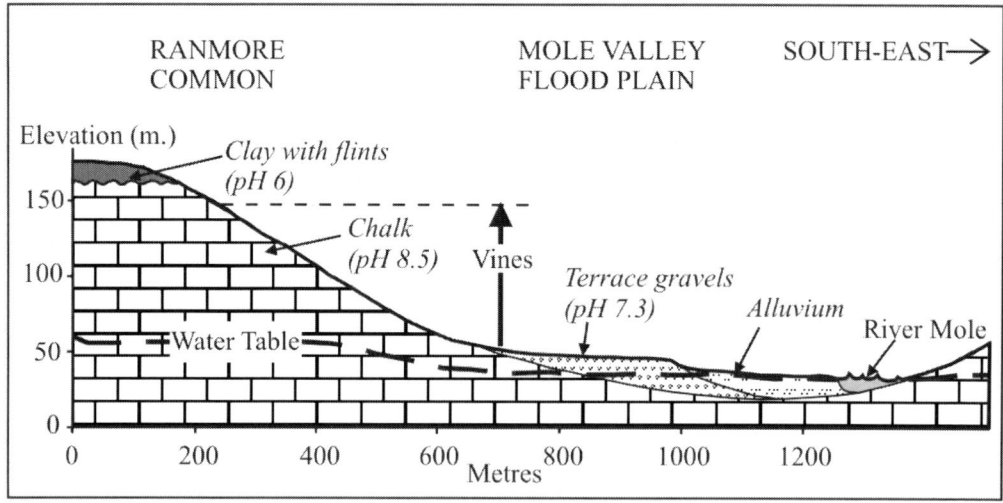

Figure 6.4 *Geological cross-section through the North Downs at Denbies Vineyard, Surrey. Vines have been planted on the Chalk slopes from an altitude of 145m down to the glacial River Terrace gravels of the River Mole at 45m.*

Indeed the place that he considered particularly suited was 'The Dell', a dry valley protected by the headland (Fig. 6.5). Adrian White acted on this suggestion and, with the help of German viticulturalists from Trier and Giesenheim, in 1983 changed Bradley Farm into a vineyard. 35 acres were planted in 1986, 30 in 1987, 70 in 1988 and a further 65 in 1989. Finally some 260 acres had been planted with 19 different varieties of grape (Fig. 6.6). About half of the vineyard occupies sloping Chalk. The rest is on 'brick earth', and river terrace gravels of the River Mole. Brick earth is comparable to 'loess', a wind blown silt, while the gravels contain fossilised mammoths and other denizens of the Ice Age.

Figure 6.5 *General view of Denbies vineyard seen from Box Hill to the east. The vineyard stretches from the periglacial dry valley (the gap between the distant trees) down the chalk slopes and across the raised river terrace of the River Mole in the foreground. The winery is hidden by the right foreground foliage (see front cover). In the far left background are the abandoned winelands of the Surrey Hills (Lower Greensand).*

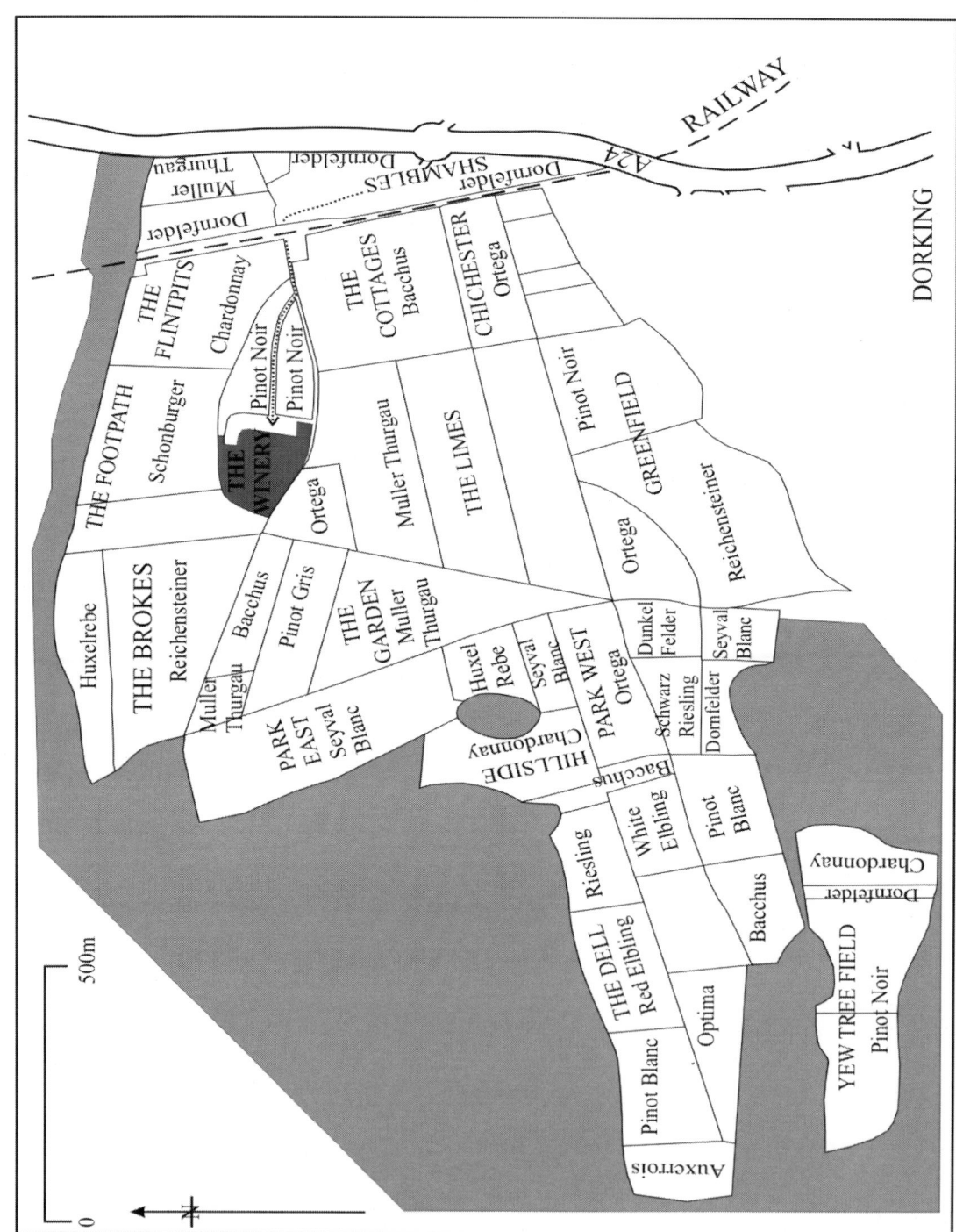

The Terrace Gravels are flat and well-drained being slightly higher than the modern alluvial flood plain of the River Mole (The origin of terrace gravels and their significance to viticulture is discussed in the next chapter.).

Not that the metamorphism of a brambly pig farm into a vineyard was done without much opposition from the local (Mole Valley District) Council, stimulated into action by the aboriginal inhabitants of the area. Oblivious to the history of viticulture in the neighbourhood, plans submitted to the local council for the vineyard were opposed every step of the way - under the banner of 'abuse of aboriginal rights'. There was concern that the aromatic old pig farm would be replaced by industrial development, and that tired and emotional wine critics would shatter the rustic calm of nearby Dorking. The building that now houses the winery, restaurant and visitor centre has received an architectural award, and Denbies was voted SE England's top tourist attraction in 2002 (Fig. 6.6).

Ten years since it was established Denbies produces some 200,000 bottles of wine a year. This includes a range of white wines, including *méthode champenoise* sparkling wine, red wines produced by malolactic fermentation, rosé and, when the season permits, *Botrytis* dessert wine. Because of the high quality but small quantity of the red wine, plans are underway (2003) to replace some of the white wine vines with Rondo, a grape for producing red wine that originated in Manchuria, as mentioned above.

6.3 ENGLISH (AND WELSH) WINE

Until the last 20 years of the 20th century wine drinking was very much the preserve of the gentry. Beer, cider and spirits were the norm. Wine, imported mainly from the European mainland, was a high status drink, surrounded by arcane language and rituals. This was a return to the days before the Roman Conquest, when wine was the high-status drink of a few Celtic Chieftains and their aspirational cronies. If drinking imported wine was for the few, drinking English wine was for the very few. Few people were aware of the two millennia of tradition of English viticulture. This was true, not only of the general public, but even professional wine-writers were ignorant of this history. Thus one wine critic, who shall remain anonymous, recently wrote 'The English wine industry started again in earnest in the 1960s after a pause of some 400 years'.

This book is concerned with describing the impact of climate and geology on British vineyards, not the resultant wine. Nonetheless the opportunity cannot be lost of establishing the credentials of English wine to non-believers. The new English vineyards were planted with varieties of largely German origin, and vignerons, or vinearons, to use Hyams created word, were producing Germanic style wines. They were light, floral, but often highly acidic 'pucker-up' wines. French critics would say that one could taste the rain in them.

Half a century after the renaissance of English viticulture there has been a renaissance in English vinification. Over the last decade there has been a big palate shift in the style of English wines.

Figure 6.7 (Facing page) Map of Denbies vineyard, Surrey, showing the distribution of areas planted with 20 different vine varieties in 1990. Note how many of the grape varieties have been planted both on the Chalk slopes (Left), and on the river terrace gravels (Right). Subsequently many of the white wine vines in the Dell have been removed and replaced by Rondo, a vine for making red wine.

New World vinearons, and Britons who have trained in the New World, have brought high-tech. vinification to English wineries. This has resulted in a major change in the characteristics of English wines. Red wines in marginal cold climates, such as England, tend to have high acidity. Many of the early English red wines were memorably awful. The initial high acidity can be counteracted, however, by malolactic fermentation. This is being applied to English red wines, which are now, in that congenial phrase, drinking well. Even so red wine makes up only about 10% of English wine production.

English white wines have changed dramatically in character too. Many are now being oaked. Others are subjected to the *méthode Champenoise*, producing zesty sparkling wines, to which the name of Champagne may not, of course, be applied. As an aside, it is not generally appreciated that Champagne was discovered by the English, and invented by the French (Simon, 1962). In former times the filtration of yeast from wine was not as effective as today. Wine shipped to England in the autumn would rest in cool cellars. When the wine warmed up, either in spring, or because it was brought up into the house, the yeast would be reactivated and the wine underwent a secondary fermentation. The carbon dioxide gas would remain in solution until the cork was removed from the bottle by corkscrew or high pressure. The English liked the bubbly wine, though drinking it was a high-risk sport due to the weak glass and resultant exploding bottles, together with occasional ricocheting corks. Champagne drinking became safer, though less exciting, with the advent of stronger bottle glass and improved immensely when the blessed Madame Veuve Cliquot developed the *méthode Champenoise* in the early 19th century.

The English climate is also suitable for the occasional and eccentric production of dessert wines. In autumn grapes may be attacked by the fungus *Botrytis cinerea*. This thrives on grapes in damp misty autumn mornings. If the day continues damp and cloudy the fungus will reproduce with such abandon that the grape skins expand and split, the dreaded 'grey rot' will set in, and the crop be ruined. If a damp morning is followed by a warm afternoon, however, the fungus will reproduce more slowly, the grapes will lose moisture and shrivel up to produce the so-called 'noble rot'. The fungus feeds on sugar in the grape, but also, and more importantly, removes acidity. Though some sugar is lost, the overall effect is to produce a shrivelled up grape with high sugar content. The resultant dessert wines are akin to those of Sauterne and Hungarian Tokai. By some they are highly prized. The English climate occasionally has autumns that are favourable for the 'noble rot' to set in. Being rare and labour intensive to pick, such wines are very expensive. Fortunately this writer finds that they have a taste akin to gone off sherry. This last point is perhaps the final one to bear in mind when considering English wine. Some people prefer French wine, others German, others Californian. What does it matter? It is no more sensible to rank wines of different countries, than to rank bananas, apples and pears. Wines differ widely from one another, not only from within the same country, but from within the same region and within the same vineyard, and from year to year. One should drink what one enjoys, not what one is told to enjoy. Taste is a personal matter. *Vive la difference!*

6.4 REFERENCES

Barty-King, H. 1977. A Tradition of English wine. Oxford Illustrated Press. Oxford. 250pp.

Johnson, H. 2004. Pocket Wine Book. Mitchell Beazley. London. 288pp.

Selley, R.C. 1996. The Geological Inspiration for the Denbies Vineyard. Denbies Wine Estate. Dorking. 8pp.

Simon, A.L. 1962. The History of Champagne. Ebury Press. London

Skelton, S. 2001. The Wines of Britain and Ireland. Faber & Faber. London. 531pp.

Tate, A.B. 2001. Global Warming's Impact on Wine. Jl. Wine Research. 12. 95-109

Unwin, T. 1991. Wine and the Vine. An Historical Geography of Viticulture and the Wine Trade. Routledge. London. 409pp.

Mann, R. 2002. UK Wine: A basic guide for consumers. Gateway Books International. Singapore.117pp.

THE WINELANDS OF BRITAIN

PART 3.

BRITISH WINELANDS:

PROSPECTIVE

CHAPTER 7.

The future of British viticulture in a changing climate

I fear tundra in Tonbridge, and glaciers in Guildford more than vineyards in Kent.....
Letter in the TIMES 1 November 2001

It is a matter of great current concern whether the Earth's climate is static, warming or cooling, and whether climate change is harmful or beneficial to civilisation. The impact of climate change on viticulture is of particular interest to wine lovers. This interest is not new. Some 2000 years ago Saserna (cited in Columella, 67AD) attributed the northward migration of viticulture across the Roman Empire to global warming (*plus ça change, plus c'est la même chose*).

7.1 CLIMATE CHANGE: EVIDENCE AND CAUSES

The geological record provides ample evidence of climate change extending back in time for at least a thousand million years. Sedimentary rocks, and the plant and animal fossils that they contain, indicate deposition in a range of climates from glacial to hot dry deserts. In the British Isles for instance, rocks of glacial, hot humid and desert conditions occur interbedded with one another, albeit with intervals of many millions of years separating strata laid down in different climates. The juxtaposition of rocks laid down in different climates on the same small island may be explained by continents drifting from equatorial latitudes to polar latitudes, and continuing over the pole to the tropics again. There is also, however, evidence of global climate change, with the Earth alternatively experiencing uniform hot humid and cold arid climates. These are termed 'greenhouse' and 'icehouse' episodes, the latter enjoying the media-friendly synonym 'snowball Earth' event (Walker, 2002).

In addition to sedimentary and fossil data, there are now various techniques for calculating past temperatures from the geochemical analysis of rocks. Crosschecking all these different palaeo-thermometers enables them to be integrated to produce a graph showing how the temperature of the Earth has varied though time (Fig. 7.1). The accuracy of this graph is rather like the human memory. It is highly reliable for recent events, but becomes less so going back in time. The actual amount of temperature change, and the rate and frequency at which changes occur become progressively 'fuzzier' with time (Hardy, 2003).

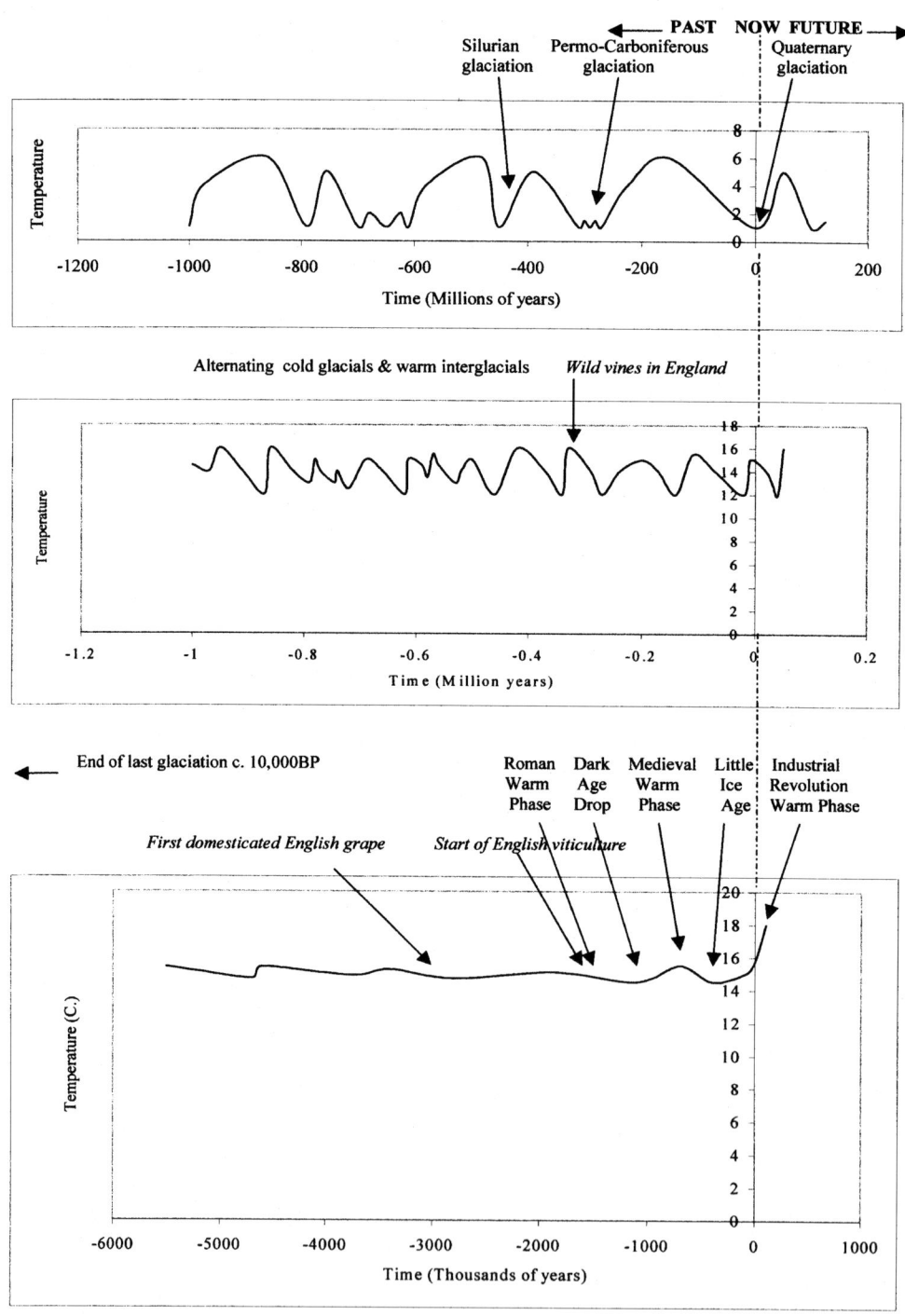

In general terms the planet has been cooling down for the last 60 million years or so. That is from the time of the extinction of the dinosaurs and the Chicxulub impact. The two events may have been connected, but if they were, the dinosaurs had remarkable pre-cognition. They, and many other species, were already in decline for millions of years before the impact. This gradual global cooling has plunged the planet into another ice age for the last two million years. There have, however, been alternating cold (glacial) and warm (interglacial) periods (Siegert, 2001). During the glacial maxima most of northern Europe was covered by ice. At its maximum the polar ice cap extended as far south as the Thames Valley. The southern Home Counties enjoyed tundra conditions with herds of hairy mammoths and reindeer. During the warm inter-glacials, by contrast, the Thames Valley was the haunt of hippopotami, hyenas and hominids. The most remarkable feature of the climate curve is the thermal stability of the last ten thousand years, an almost unique event in the record. Experience dictates that a climatic fluctuation is imminent; the question is, in which direction. Since we are now in the middle of an Ice Age, albeit enjoying a warm interglacial, it is reasonable to suppose that the next change will be a drop in temperature, and the return of another glacial maximum. This has been the fear of climatologists for half a century or more (Manley, 1952).

Examination of Figure 7.1 shows that there have been several fluctuations in temperature within the last two millennia. There was a warm spell through the Roman and Medieval periods, though with an intervening Saxon sag, when the temperature was 0.3^0C above a reference point at 1900. During the ensuing Little Ice Age from the mid-15th – mid 19th centuries temperature dropped by nearly a whole degree to 0.6^0C below the 1900 benchmark. Temperature began to rise again in the middle of the 19th century at the commencement of the Industrial Revolution, and has continued upward ever since. There is evidence that bud burst is starting earlier on mainland Europe, as much as two weeks in Austria (Brunlmayer, in Payne, 2002). Data on the start of the '*Vendage*' in France are inconclusive (Ladurie, 1983). With all the panic about global warming and its impact on British fauna and flora, it should be noted that the temperature has only just got back to where it was in Roman and Medieval times.

The prevailing expert opinion is that the earth is now entering a phase of rapid warming. The actual rate of warming varies according to the computer models being employed (Watson et al. 2001). Many climatologists believe that this warming is caused by an increase of carbon dioxide in the atmosphere. Many further believe that this results from the combustion of fossil fuels since the start of the Industrial Revolution in the middle of the 19th century.

Figure 7.1 (Facing page) Graphs to show climate change during the recent history of the Earth. For each graph temperature increases vertically. Absolute values have only been measured since 1660. Phenological observations, and tree ring studies can establish relative values of temperature further back, as can isotope analysis, to some 70 million years B.P. (before present). For more distant time climate is estimated less robustly still using geological evidence, such as ancient glacial (cold), and salt (hot) deposits. Upper: graph from 1,000 My BP showing alternation of 'ice house' and 'green house' episodes. Prediction of future 'green house' episode is based on the periodicity of the galactic year. Middle: graph for the last 10 My BP, based on isotope data from polar ice and ocean floor sediments. Past periodicity suggests that another glacial advance is imminent. Note that wild vines grew in England during previous warm interglacials, as well as the present one. Lower: graph for the last 5,500 years. Note that the temperature has only just climbed back to that enjoyed in Roman times. Most climate models predict future rising temperature. For sources see the Nansen Arctic Drilling Program Science Committee (1992), Kroon, D.,Norris, R.D., Klaus, A., and the ODP Leg 171B Shipboard Scientific Party (1988) Hart & Hart (2000), Skinner, B.J. & Porter, S. C. (1995), French (2002), Steiner (1967), Veevers (1990), Imbrie & Imbrie (1979), Merritts et al. (1998), Cuffey and Brook (2000) and Lamb (1995).

This view is by no means universal. Some astronomers hold that climate change is due to cyclic astronomical variables, collectively known as Milankovitch cycles, named after the eponymous professor. These variables include the precession of the equinoxes, the obliquity of the earth's orbit and the orbital eccentricity (Fig. 7.2).

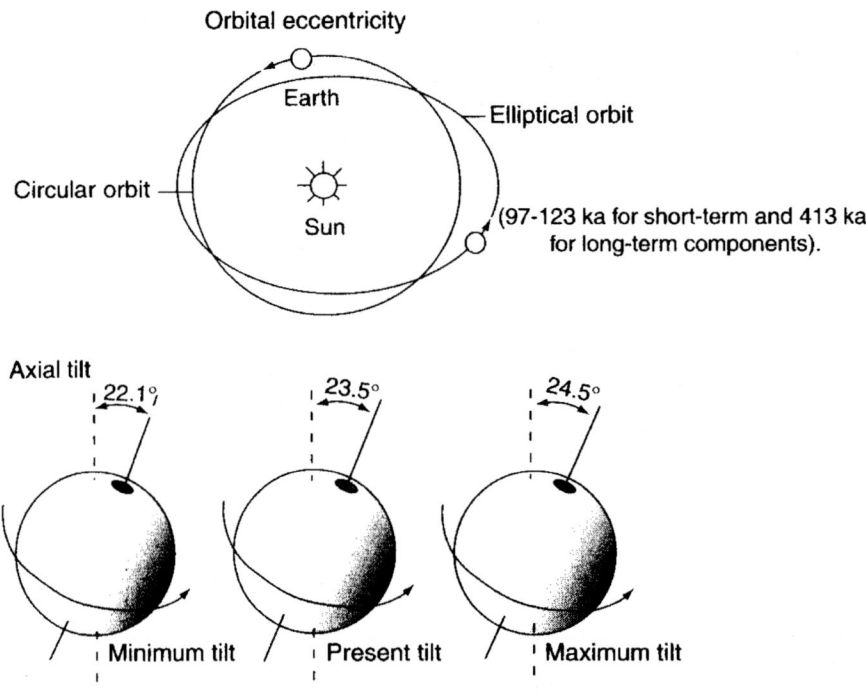

obliquity of the Earth's rotation axis (41-54 ka)

Figure 7.2 Illustration of the Milankovitch cycles due to the interplay of the eccentricity, obliquity and precessional cycles of the earth. These are believed to affect global climate. The integration of their periodicity generates apparently random climatic changes. These can, however, be unravelled with the help of some cunning statistical gymnastics. (From Kroon, D., Norris, R.D. Klauss, A., and the ODP Leg 171B Shipboard Scientific Party, 1998. Drilling Blake Nose: The Search for evidence of extreme Palaeogene-Cretaceous climates and extraterrestrial events. Geol. Today. 14. 222-225. Fig. 4, Courtesy of Blackwell Science.)

The interplay of these variables can generate changes in climate that superficially appear to be random, though statistical gymnastics can differentiate the periodicity of the original signals (This periodicity is best seen in the middle graph of Fig. 7.1 that shows how the QuaternaryIce Age within which we live, consists of alternating glacial and interglacial episodes). This graph also suggests that renewed glaciation is imminent, on a geological time scale, at least. Glaciation may be more to be feared than global warming. Supporters for astronomically driven climate change argue that the next glaciation may be postponed by the combustion of fossil fuels, and that we should keep the home fires burning. This is not the place for a detailed critique of the arguments pertaining to environmental change in general and climate change in particular.

The onset of a new glaciation spells doom for the human race, but then so perhaps does the onset of rapid global warming. Since Saserna (cited in Columella in AD67) there have been speculations on the influence of climate change on viticulture (Jones and Davis, 2000). Tate (2001) has presented a detailed scenario of the effect of global warming on viticulture. He analysed the effect of global warming in general, and of the accelerating melting of the Arctic ice cap in particular. Global warming will cause a general shift of wine-growing areas to lower latitudes in both hemispheres, broadly between the 10^0 and 20^0 C. isotherms. The equatorial margins of winelands will be abandoned, and their polar margins migrate poleward. Not only will global warming affect the overall extent of viticulture, but there will be smaller scale changes too. Pinot Noir in the western USA and Europe may gradually migrate northwards, while Burgundy may become too hot for this vine variety to flourish. Not only wine varieties, but also wine diseases, will also migrate poleward. The effects of global warming on Europe are, however, more complex. It has been argued that the melt waters of the Polar ice cap will flow past northwest Europe diverting the Gulf Stream to the south and inhibiting excessive heating of the land. Either way, thermal expansion of the oceans coupled with the melting of the polar ice caps will result in rising sea level. A rise of 0.5m per century is currently predicted. Given the presently accelerating rate of melting of the Polar ice cap this may be rather low. Sea level rose 5m per century at the end of the last glaciation (Hanebuth *et al.* 2000). As the Gironde estuary floods the vineyards of Bordeaux will drown.

Even the (diminishing) hole in the ozone layer may not be all that bad. It is believed to encourage Australian and New Zealand vines to produce poly-phenols in their skins, enhancing flavour and quality (Smart, in Payne, 2002). The stability of the last 10,000 years is unlikely to continue, and whichever way the climate changes, the short-term future of viticulture in Britain looks bright, and we should enjoy our liquid assets while we may.

7.2 BRITISH VITICULTURE AND PAST CLIMATE CHANGE

It is interesting to review British viticulture in the context of changing climate, and it will have been noted that the historic chapters of British viticulture in this book have been arranged according to discrete climatic episodes. There is no evidence of viticulture in pre-Roman Britain. The Romans introduced viticulture into the island. The archaeological data presented in Chapter 2, demonstrate that viticulture was not just the hobby of homesick Roman expatriates, but a commercial undertaking that was carried out even in the north of the province. Thereafter, though the archaeological and literary evidence is patchy, viticulture has continued in Britain uninterrupted for the last two millennia, a fact of which the general public is largely unaware. Viticulture seems to have declined in the Dark Ages, though the decline may be more apparent than real, due to the paucity of preserved literature. A renaissance begun by the Normans is well documented in the Domesday Book. Both the number, and also the size of the new vineyards shows that they were not being planted for sentimental reasons by homesick Norman knights, but with a view to substantial commercial wine production. The decline of viticulture in the 14th century is generally attributed to the Black Death and other socio-economic factors, but the coincidence of the decline with the advent of the Little Ice Age points to climate change as another factor.

Likewise it is interesting to speculate on why the 20th century renaissance of viticulture was concomitant with the rise of the Industrial Revolution, and global warming. Table 7.1 documents all known vineyards in Britain, including, Wales, the Channel Islands, and, improbably, Scotland for the last two thousand years, and figure 7.3 shows the location of all the vineyards in the Roman, Medieval, Little Ice Age and Industrial Warm Phase maps from previous chapters.

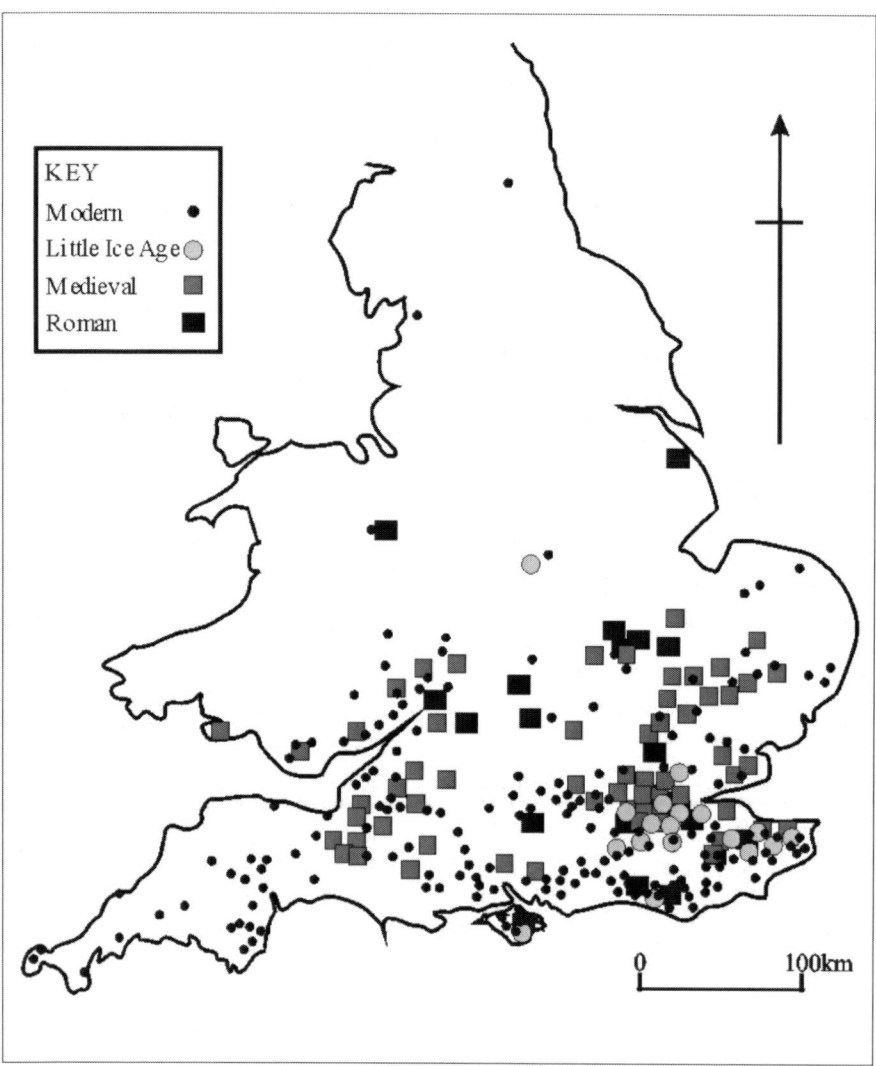

Figure 7.3 *Map of Britain showing the location of vineyards of the Roman, Medieval, Little Ice Age and Industrial Warm Phase maps from chapters 2, 3, 4, and 6 respectively. Note how Little Ice Age vineyards were few, and largely restricted to the southeast of England, whereas modern vineyards now re-occupy the Roman and Medieval winelands, southeast of the Humber – Severn line.*

Table 7.1 *(Facing page) British vineyards through the ages. This subsumes data largely presented in the tables in the earlier chapters.*

CENTURIES:	2-4th	10th	11th	12th	13th	14th	15th	16th	17th	18th	19th	20th	21st
ENGLAND													
Beds.			1	1		1						1	
Berks.			3	3	4							6	6
Bucks.	1		1										6
Cambs.	1		1	1	3							3	4
Cornwall					2			1		2		1	8
Derby								3				1	
Devon						2						4	11
Dorset			2	1								1	5
Durham				1									
Essex			10		11	1	2	3			1	12	15
Gloucs.	2		1	1		2	1	1		5	2	1	8
Hants.	1		1	1						1		8	21
H'ford.				3	2					2	1	3	7
Herts.	1		3		2				2	1		1	6
Isle of Wight	1									1		3	4
Kent	2		3		11	11	2	4	2	2	1	11	33
Lancs.													1
Leics.												1	4
Lincs.	1			1								2	1
London	3				10				3	17	1		2
Middx.			8										
Norfolk												9	2
N'hants.	3				1							1	3
Notts.										1		1	1
Oxon.	1								1			2	7
Rutland													1
Salop.										1			3
Somerset		2	7	1	3	4						7	15
Staffs.									1			1	1
Suffolk			3		2							10	12
Surrey	2			1	1	1			3	2		4	8
Sussex (E.)	2				2			1				13	24
Sussex (W.)					2					1		4	7
Wilts.			4									6	6
Worcs.				4	2			1		2			4
Yorks.													3
WALES				3							1	6	12
CHANNEL ISLES												1	2
SCOTLAND				1									
TOTALS:	19	2	48	23	58	21	6	14	12	36	8	124	251

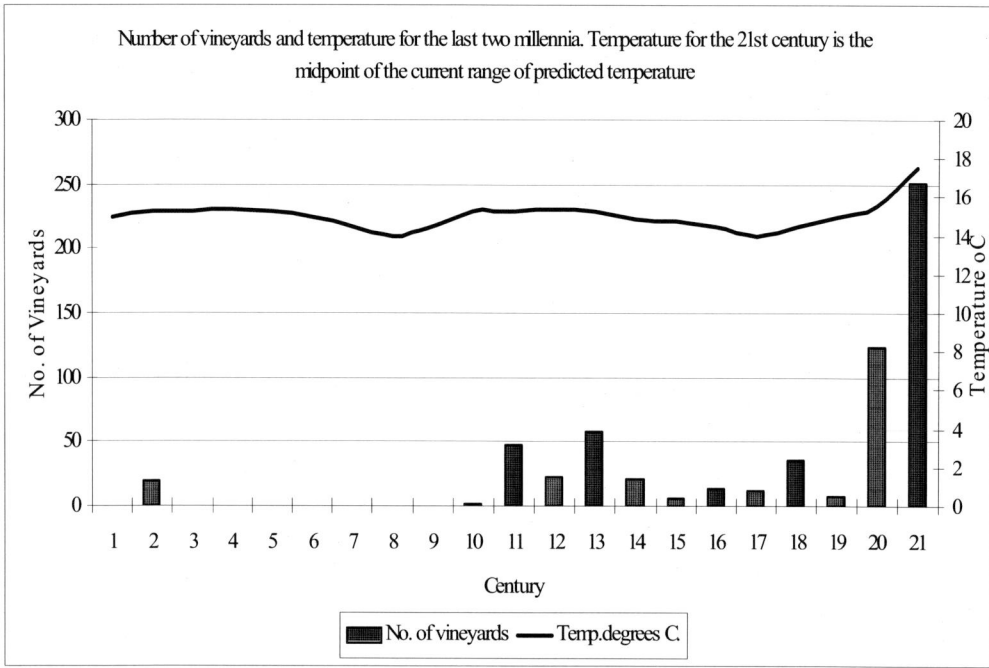

Figure 7.4 *Histogram showing the number of vineyards in Britain over the last two millennia (Data from Table 7.1), coupled with a graph of temperature compiled from data in Houghton and Jenkins (1990), Skinner et al. (1995) and Watson et al. (2001). The temperature for the 21st century is the mean of the range of models presently calculated. The problems of data reliability and caveats against concluding that there is an obvious causal relationship between temperature and the number of vineyards are discussed in the text.*

Numbers of known vineyards, century by century are plotted against temperature in Fig. 7.4 above. Superficial examination of this figure suggests a broad correlation between temperature and the number of British vineyards. It may be tempting to believe that the link is causal. It is important to step back, however, and to carefully consider both the reliability of the data and the apparent correlation. Consider first the data quality, both thermal and the numbers of vineyards, century by century. The data quality degenerate going back in time for both of these variables. Accurate temperature measurements only commenced towards the end of the seventeenth century. Data quality drops off sharply before then. The numbers of vineyards are accurately known for the 20th and 21st century. Prior to then the only robust benchmark is provided by the Domesday Book of 1087. The number of vineyards given for intervening centuries should always be considered as 'greater than' the number recorded, since the information, largely literary, must be incomplete. In particular data for the Dark Ages (5th – 10th centuries) are scarce, either because there were no written records, or because they have been lost. Similarly both the number and ages of Roman vineyards is very 'soft'. In Figure 7.5 they have all been grouped together in the 2nd century, though they may have flourished from the 2nd to the 5th centuries. Roman vineyards have been identified from data of a wide range of reliability, ranging from 'terracettes' and grape pips, of dubious worth, to extensively excavated extensive sites.

Having critically reviewed the reliability of the data, it is appropriate to consider the relationship between temperature and the number of vineyards. A cursory look at Figure 7.4 suggests a rough correlation, particularly from the Little Ice Age to the 21st century. This is, however, the application of the principle of simplicity, termed Occam's Razor, after the eponymous 13th century monk, also known today as the KISS principle (Keep It Simple Stupid). The simple view that temperature and number of vineyards are causally related is a dangerous conclusion to reach, akin to concluding that TV drives people mad, because admissions to mental institutions directly correlate with sales of TV sets. Both correlate with increasing population, they are not causally linked – are they? It has been remarked that geologists often use statistics like a drunk uses a lamppost; more for support than illumination.

The data for Britain are too soft to justify the conclusion that viticulture correlated with temperature, but. Climatic fluctuations cause alterations in the environment that drive socio-economic change including, and inextricably linked with, plant and animal life. Variation of one parameter feeds back into another. This interrelationship is what geologists call Earth system science (Jacobson et al. 2000), and what Lovelock calls Gaia (Lovelock and Margolis, 1974), named after the ancient Greek goddess of the earth, the archetypal earth mother.

Throughout the two millennia during which viticulture has been practiced in Britain it is apparent that geology, as ever, underpins all activity. Throughout climatic fluctuations well-drained south-facing slopes have been the preferred, though not exclusive, location. It is interesting to note how vineyards have been planted, decayed, and then replanted on the same site for geological reasons. Examples include Painshill Park, Cobham, Surrey, first planted in the 18th century, Toppesfield, Essex, recorded in the Domesday Book, and Pilton, Somerset, planted in the 13th century. All of these were replanted in the 20th century.

7.3 THE RELATIONSHIP BETWEEN BRITISH VINEYARDS, SCENERY AND UNDERLYING GEOLOGY

Before proceeding to speculate on the future of British viticulture in an era of changing climate it is essential to have a basic understanding of British geology. First a simple geological principle needs to be understood: the Law of Superposition of Strata. This sounds frighteningly learned, but is actually very simple. High rise buildings in the City of London have deep foundations cut down through layers of interbedded mud, sand and gravel. With increasing depth of excavation strata contain artefacts and detritus from the blitz of World War II, then of the fire of London, then of the Roman occupation. These observations demonstrate the fundamental geological principle of the superposition of strata – younger strata overly older strata. Deeper excavations reveal layers of sediment that contain prehistoric remains. These include the tools and bones of hominids, and the bones of other animals that now live in climates far different from that of today, warmer in some strata, cooler in others. The alternation of these tropical and glacial fossil assemblages demonstrates the cyclicity of climate change.

Beneath these sediments lies the London Clay of Eocene age (deposited some 55 – 34 million years ago) with fossils of animals and plants, including vines, though of species different from those of today. London lies in a large basin of strata (syncline). The Cretaceous Chalk limestone crops out south of London in the North Downs, and north of London in the Chiltern Hills (Figure 7.5). The chalk contains fossil fish, bivalves, and other invertebrates that look stranger than those of the London Clay, and still stranger than those of today, together with extinct fossils of groups such as the ammonites.

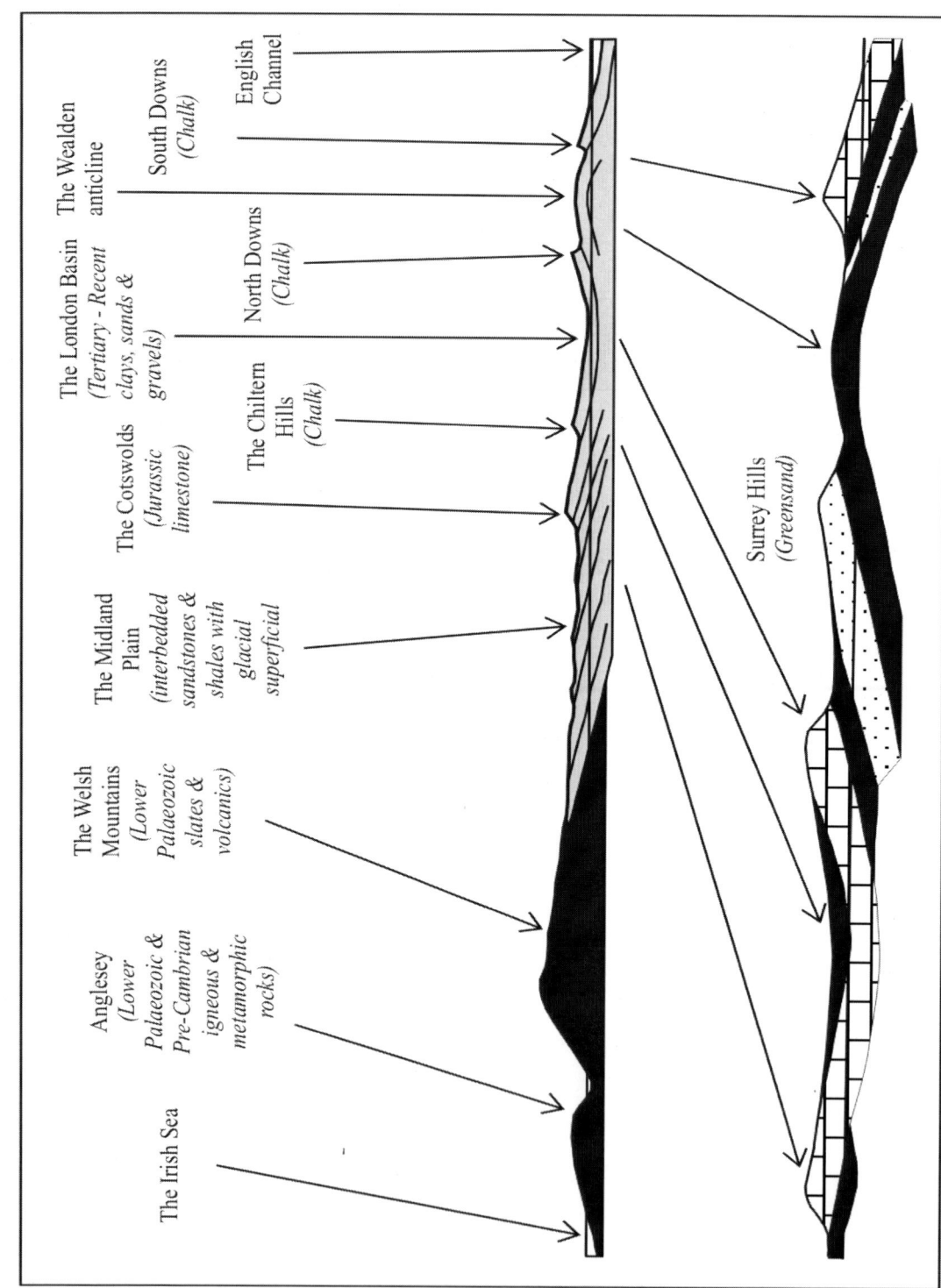

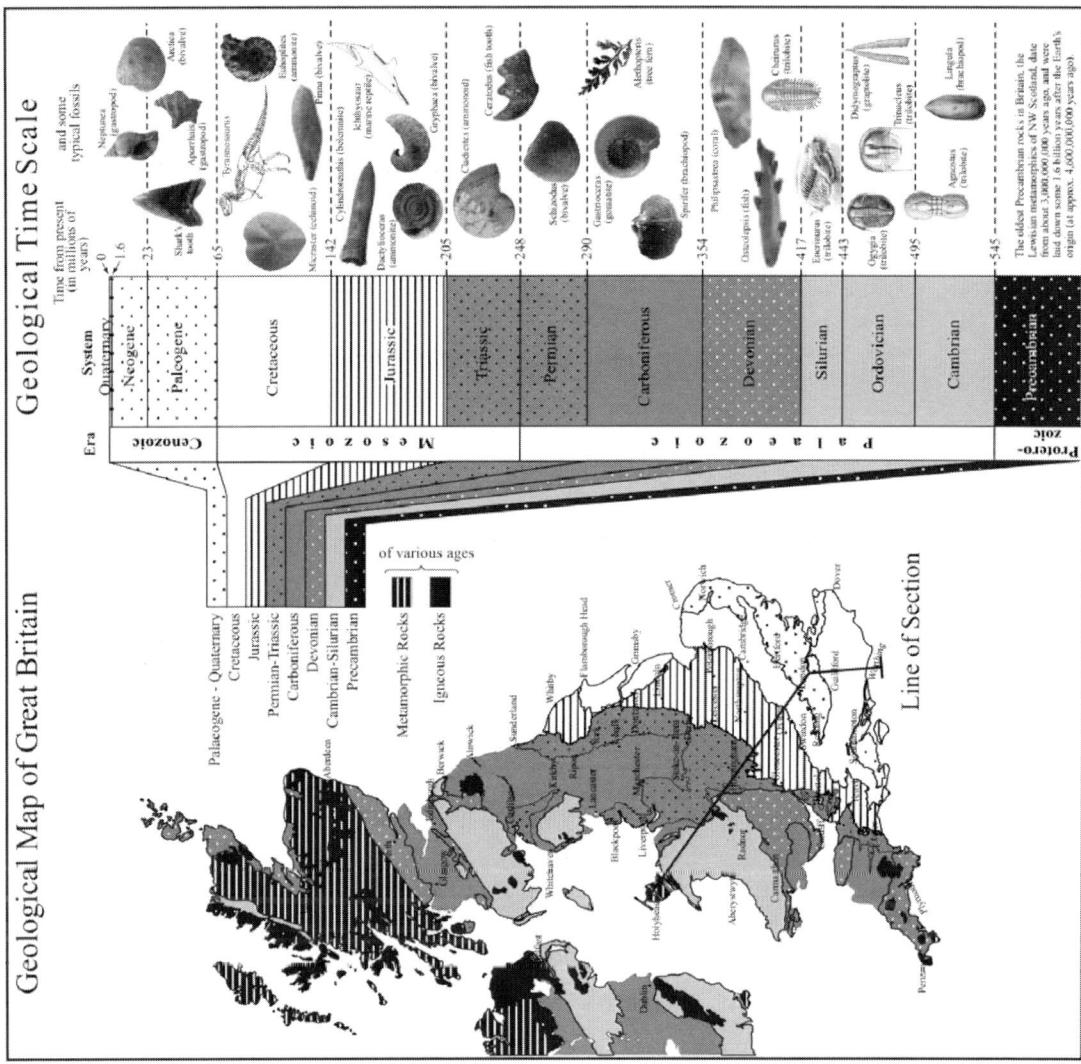

Figure 7.5 (Above) Geological *cross-section from Anglesey south-east to the English Channel. Location of the section is shown in Figure 7.6. The lower cross-section shows more detail of south-east England. Both cross-sections show the relationship between geology and scenery. Note in particular how the eroded edges of folded strata give rise to unfavourable shady north-facing scarps and sunny south-facing scarps, depending on the orientation and dip direction of the strata.*

Figure 7.6 Right: *Geological map of Great Britain and Geological Time Scale, courtesy of the School of Earth Sciences and Geography, Keele University. In simple terms the rocks that crop out at the surface increase in age from southeastern England to the north and west, though this generalisation is complicated by the gently folded nature of the strata in the southeast. For an enlarged and brightly-coloured version of this map see the Frontispiece.*

85

As shown in figures 7.5 and 7.6 traversing north and west from the London Basin one descends a series of escarpments, encountering progressively older and older strata (remember the Law of the Superposition of Strata). These strata were first mapped out and their contained fossils described in the early 19th century by William Smith, known as the 'father of English geology'. In 1815 Smith published a famous geological map of Britain accompanied by cross-sections through the strata (Winchester, 2001). Descending the Chilterns one moves across a series of clay bottomed valleys and resistant ridges, the most prominent of which is the Cotswolds. The limestones of the Cotswolds are made up of the smashed (but rarely entire and collectable) remains of ammonites, corals, crinoids and other invertebrates. The Cotswold limestones in turn overly the Liassic shales that crop out from Whitby in Yorkshire to Lyme Regis in Dorset. The Lias is famous for its extinct marine vertebrates, such as the *Ichthyosaurus* and *Plesiosaurus*, as well as ammonites and other fossils. Below the Lias is the New Red Sandstone, largely barren, except for rare reptilian bones and tracks. The New Red Sandstone in turn overlies the Coal Measures, with their horizons rich in extinct plants, and beneath the Coal Measures is the Carboniferous Limestone, with fossils of extinct species of coral, crinoids, brachiopods and other marine invertebrates. The underlying Old Red Sandstone is largely barren, like the New Red, but yields rare fossil fish. Many of these are really weird, with bony external skeletons. By now we have reached the mountains of Wales, in whose slates are found no mammals, no reptiles, and indeed no vertebrates at all, but only fossils of trilobites, the strange pencil mark palimpsests of graptolites, unfamiliar brachiopods and gastropods. In Anglesey the slates rest unconformably on unfossiliferous igneous and metamorphic rocks (A summary stratigraphic column was illustrated in figure 5.7).

Readers who doubt the veracity of the previous brief account may study the maps and memoirs of the British Geological Survey, or textbooks on British geology. Out and out sceptics may prefer to confirm these observations for themselves, donning a pair of boots, and taking up a hammer, a hand lens and a notebook, as geologists have done since William Smith. In their travels they will encounter many other geologists, amateur and professional, who are confirming the evidence just summarized. During university vacations professors demonstrate this sequence of rocks and their fossil record to parties of cold, wet sceptical students.

This sequence of evolving fossils, from the first primitive plants to the invertebrates, to the appearance of fishes, then amphibians, then reptiles, then mammals, and finally hominids has been encountered all over the world. The observations are confirmed on a daily basis. This huge database demonstrates that evolution is a fact, not a theory, as 'creation scientists' would have children believe. But this is a digression from the application of geology to understanding British viticulture and climate change.

7.4 THE PROSPECTS FOR BRITISH VITICULTURE DURING GLOBAL WARMING

Given that the stable climatic conditions of the last ten thousand years are unusual and unlikely to continue, it is interesting to speculate on the future. If the climate of Britain cools, either through the advent of a new ice age, or, as some computer models predict, because global warming will melt the Arctic ice cap, whose waters will shut off the Gulf Stream, then viticulture is doomed. If, on the other hand, global warming continues, the future of British viticulture is bright, at least in the short term. Throughout the Industrial Revolution Warm Phase, vineyards have extended from southern England up to a line from the Severn estuary to the Humber.

They now reoccupy the old Roman and Medieval winelands planted before the advent of the Little Ice Age (Refer back to Figure 7.3). Already it is possible to see how geology determines the development of vineyard terrains similar to the legally-defined *Côtes* of France. Maps of the distribution of British vineyards during the different climate epochs (Figures 2.2, 3.1, 4.1 and 6.2) show how viticulture has ebbed and flowed across the country (Fig. 7.7).

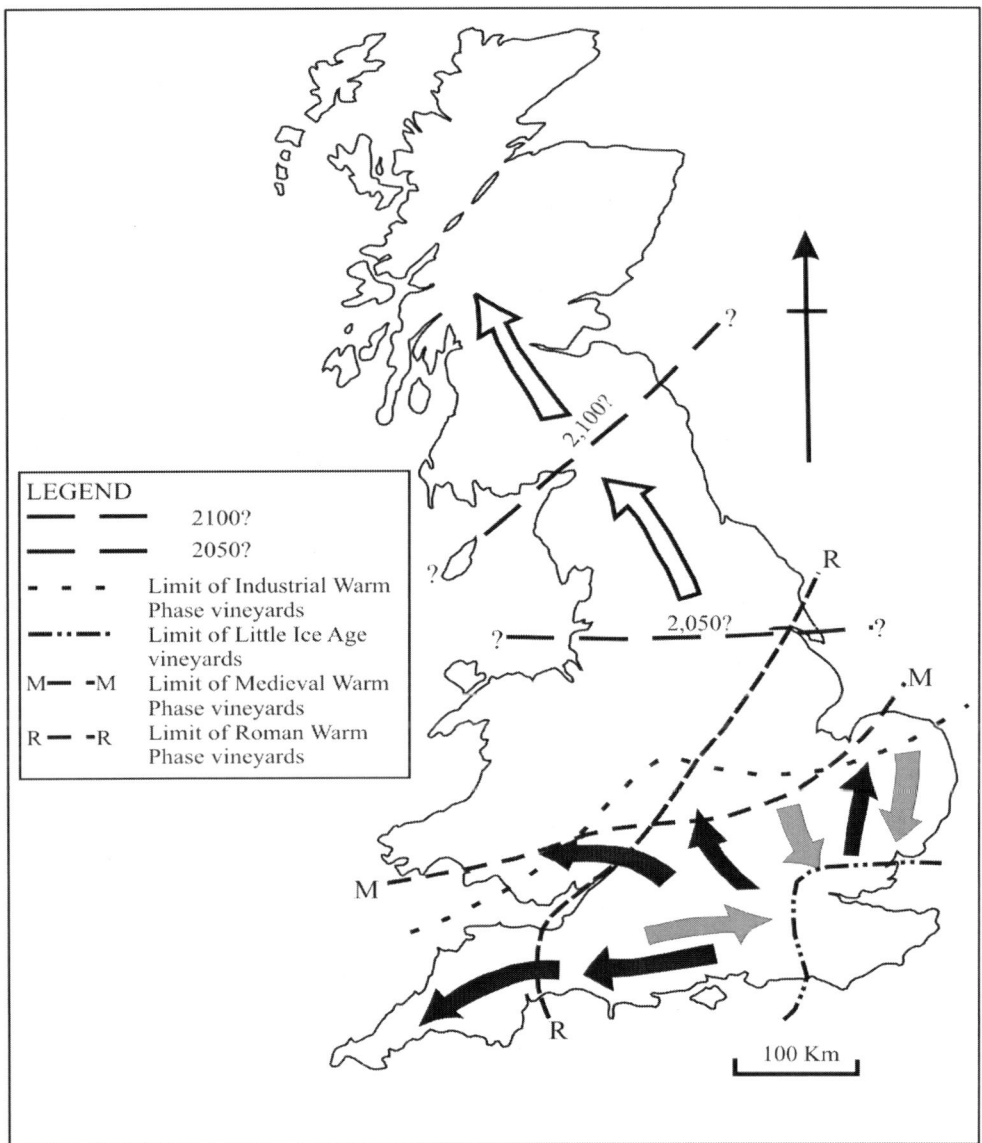

Figure 7.7 Map showing the ebb and flow of British winelands, during the past 2 millennia. After their retreat to southeast England during the 'Little Ice Age' vineyards are now established further north than in Roman and Medieval times. If global warming continues the northward advance will extend to virginal winelands in northern England and then Scotland.

The distribution is irregular, however, because vineyards have only been planted where geology is favourable. Integrating geology and climate change allows the recognition of a number of what the French might term *'Pays'* or *'Côtes'*. The English term 'Terrain' would do very nicely, but one hesitates to use it for fear of confusion with *'Terroir'* and a directive from the EU terroiristes in Brussels. Perhaps the term 'winelands', used in the New World, is the best choice. Consideration of geology and the resultant landscape and soil make it possible to recognise 4 winelands, viz.: ancient winelands that have been abandoned and never replanted, ancient winelands that have been abandoned and replanted, and virginal (Industrial Warm Phase) winelands. In Franglais these may be termed *'Côtes ancienne abandonné', 'Côtes ancienne renaissance,* and *'Côtes virginale'* respectively. Franglais is a language fit for this purpose since it originated in the Norman-French of the Medieval Warm Period, when English viticulture was at its zenith. One of the most delightful examples of Franglais is in a medieval law report cited by Denning (1992): *'Il jetted un brickbat a le juge que narrowly missd'*. The fate of the jetter's hand was Islamic.

Consideration of geology and the resultant landscape and soil also make it possible to speculate on where vineyards may be established in the future if global warming continues (*'Côtes de futur'?*). These four categories of wineland are listed in Table 7.2, and will now be defined and discussed.

WINELAND	TOPOGRAPHY	GEOLOGY
1. Abandoned winelands		
Greensand Hills of Surrey	S. facing scarps	Greensand (Lower Cretaceous)
2. Renaissance winelands		
2.1 The Thames Valley	Raised flat river terraces	Alluvium (Pleistocene & Recent)
2.2 The English Champagne	Plains, dry valleys & scarps	Chalk (Upper Cretaceous)
2.3 S. shores of the Severn	S. slopes of the Mendips	Upper Palaeozoic sediments
2.4 N. shore of the Severn	S. slopes of the Welsh Mtns.	Upper Palaeozoic sediments
3. Virginal winelands		
3.1 The Central Weald	Dissected upland	Lower Cretaceous Sandstones
3.2 Cornubia	S. slopes of the moors	Upper Palaeozoic slates & granites
3.3 Midland miscellany	Undulating or mamillated	Sediments of several ages & types, locally overlain by Boulder Clay
4. Future winelands		
4.1 Northern England	S. slopes of the Peak & Lake districts	Assorted Lower Palaeozoic sedimentary & igneous rocks
Scottish winelands		
4.2 The Southern Uplands	South-facing slopes	Assorted Lower Palaeozoic sedimentary & igneous rocks
4.3 The Grampians	South-facing slopes	Dalradian metasediments
4.4 The Great Glen	N. shores of the lochs	Moine Series metamorphics

Table 7.2 *The winelands of Britain, past, present and prospective, their topography & geology.*

CHAPTER 7. THE FUTURE OF BRITISH VITICULTURE IN A CHANGING CLIMATE

1. Abandoned winelands

One minor wineland can be identified where vineyards were planted, abandoned and never replanted, despite subsequent warmer temperatures. Several vineyards were established on the Surrey Hills in the 17th and 18th centuries during the Little Ice Age. They were planted on the northern limb of the Wealden anticline on the south-facing escarpment of the Lower Greensand (*Côtes de sable vert?*). These vineyards were described in Chapter 4. Deepdene vineyard was typical of this type. It is interesting to speculate on why this area has not been redeveloped. Archaeological evidence shows that the aboriginal woodland of the Surrey Hills was cleared for farming during the Neolithic period (Field and Cotton, 1987).

In the 17th century John Evelyn (1620-1706) diarist, and author of 'Sylva, or a Discourse on Forest-Trees, and the Propagation of Timber' (1664), was the squire of Wotton, a village embedded in the Greensand hills. Evelyn began a programme of reaforrestation on his estates that has continued to the present day. Maps of the 18th and 19th centuries and photographs taken in the 19th century show that the Surrey Hills possessed a remarkably open landscape compared with the forested hillsides of the 21st. Surprising as it may seem to those who picture Surrey as suburban sprawl, it has the largest percentage of woodland of any English county. This continuous programme of reafforestation is perhaps the explanation for the absence of a renaissance of viticulture in the Surrey Hills.

2. Renaissance winelands

The renaissance winelands are defined by their geology and history. They throve in the Roman and Medieval warm periods, were abandoned in the Little Ice Age, and are now being recolonised.

2a. The Thames Valley

Vineyards were first planted in the Thames Valley, from Oxford down to London, in the Medieval Warm Period, but were abandoned in the Little Ice Age. Today this wineland is re-established. Though geographically disseminated, these vineyards share a common geological setting. They occur on ancient alluvial sands and gravels of the River Thames. The alluvium of the modern flood plain is, of course, largely waterlogged and, though ideal for lush water meadows, is not suited to deep-rooted vines. The Thames, and most of England's river systems, however, possesses a series of elevated terraces of alluvial gravels on the margins of the modern flood plain. These terraces were cut, and alluvium deposited, when sea level was much higher. The alluvium contains stone tools and vertebrate fossils. In general the terraces become younger down towards the present flood plain. Detailed studies have been able to identify the chronology of the alluvial terraces and to correlate them with glacial phases of the Pleistocene Ice Age (Gibbard, 1985). These raised river terraces are ideal for viticulture being flat-lying but well drained. (Fig.7.8). The Thames Valley has some of the lowest rainfall and highest temperatures encountered in England and, if temperatures continue to rise, will be one of the prime winelands. With its indigenous bibulous population, the Thames Valley is ideally placed for wine tourism.

River terraces are not unique to the Thames, but are ubiquitous in British rivers. The Roman vineyards of the *Côtes de Northants*, described in Chapter 2, were planted on river terraces of the Nene Valley. The modern vineyards of Thorncroft, and part of Denbies (described earlier), lie on Pleistocene Terrace gravels of the River Mole, a tributary of the Thames.

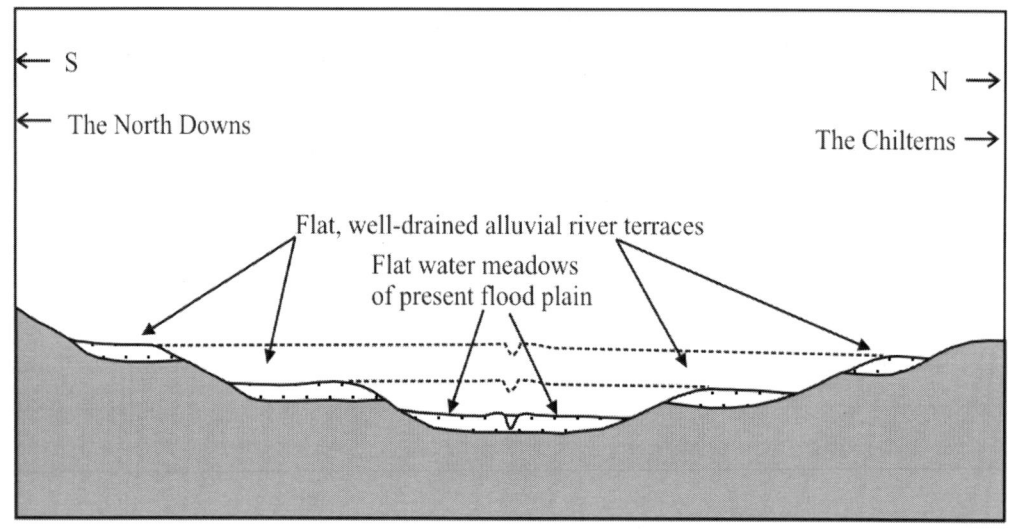

Figure 7.8 *Diagrammatic cross-section of the River Thames showing Quaternary terraces related to higher river levels. Changes in level are due to a combination of tectonic uplift and changes in global sea level, caused by the waxing and waning of ice sheets in glacial and interglacial episodes. River terraces are composed of alluvial gravel, sand and clay. Being flat and well-drained they are ideally suited to viticulture. River terraces of the Thames, and other rivers, have been widely planted with vineyards in the past and present.*

2b The English Champagne

The full weight of the French legal system will descend on anyone attempting to market 'English Champagne'. It may, however, just be possible to get away with describing the rolling Chalk Downlands as the English Champagne, using the word in its original sense of open plain, derived from the Italian *'Campania'* (From the Latin *Campus*). Chapter 5 showed how the Cretaceous limestone of the Chalk had properties peculiarly suited to viticulture. This rock is the mother of Champagne in France, and the midwife of many English vineyards. Figure 7.6 shows that the Chalk crops out in a series of tracts that radiate in every direction from Stonehenge in the middle of Salisbury Plain. Indeed the location of Stonehenge may well have been chosen because it was at the Piccadilly Circus of the ancient Downland track ways.

Vineyards have been planted on Chalk from Roman times (North Thoresby on the Lincolnshire Wolds) onwards. Hambledon vineyard, whose planting in 1951 marked the English viticultural renaissance, was located on the 'English Champagne'; many others followed. Not only does the Chalk offer the peculiar petrophysical properties suited to viticulture described in Chapter 5, but its gentle mammillated topography, dry valleys and sun-kissed south-facing escarpments offer abundant vineyard sites.

2c The southern shores of the Severn Estuary

The southern shores of the Severn estuary have been a favoured area for viticulture in Roman, Medieval and Modern times. Vineyards ancient and modern have been planted from Gloucester down to North Devon. There are several reasons for this abundance, and geology is only one of them. Certainly the role of the Severn estuary as a maritime trade route into the heart of England must have much to do with it.

The geology of the area is complex and varied. A range of rock types have been eroded into a dissected topography that provides many sheltered south-facing slopes. One particular geological feature stands out, the Mendip Hills that stretch for some 25km from Weston-super-Mare on the coast southeast to Shepton Mallet. The Mendips are composed of Carboniferous Limestone, and their southern slopes host the modern Cheddar Valley, Cufic, Rodney, Stoke Wootton, Bagborough and Avalon vineyards.

2c The northern shores of the Severn Estuary

Figure 7.3 shows another well-defined trend of vineyards ancient and modern that stretches in an arc from Worcestershire, down through Hereford, and into south Wales to extreme western Pembrokeshire. Like the southern shores of the Severn Estuary the geology is varied, but consists mainly of Carboniferous Limestones and Coal Measures and Devonian Old Red Sandstone. It is bounded to the south by the waters of the Severn Estuary and Bristol Channel, and to the north by the Cambrian Mountains and Brecon Beacons. The topography is dissected, with a regional southerly slope from the mountains to the sea. This wineland did not extend from the Roman Province of Britannia into the wilds of the Cymri, but, as described in Chapter 3, Giraldus Cambrensis (Welsh Gerald) wrote enthusiastically of the medieval vineyards of south Wales. Chapter 6 described how the renaissance of British viticulture began near Cardiff at Castle Coch.

3. Virginal winelands

Having considered the geological context of the ancient winelands reborn, it is appropriate to describe those that only appeared in the Industrial Revolution Warm Phase. Two may be recognised: the Central Weald, and the southwest of England, anciently called Cornubia.

3.1 The Central Weald

The southeastern corner of England is occupied by a huge upfold of strata (anticline) that extends across the Channel into the Pays de Brays. It is composed of sediments of Cretaceous age (geopedants will remark that there are one or two inliers of Jurassic rocks in the core). The Chalk of the North and South Downs marks the northern and southern limb of the structure, with scarps that face in towards the axis of the anticline. Inward facing scarps of Greensand are well developed on the north side, but poorly in the south because this rock formation is thin. Heavy wet sticky Weald Clay, however, occupies the major part of the Weald (Fig. 7.5). Once the Weald was covered by primeval oak forests. Because of its heavy clay clearance did not begin until the Saxons arrived with the coultered plough that could tackle such tenacious soil; hence the name of the Weald (Old English 'Wild'). There were reports of dragons in the Wealden forests right up until the 16th century. The Tudors began a major phase of deafforestation, using the timber for ships, and for charcoal. The charcoal was used for gunpowder and furnaces to smelt the iron ore that had been discovered in the Weald Clay. The Weald was thus the industrial powerhouse that supported Elizabethan and Jacobean warfare.

The Weald clay has been eroded from the core of the anticline to expose older sandstones in the High Weald of St Leonards' and Ashdown forests. These sandstones crop out in cliffed and dissected hills. Though the Weald was cleared and civilised by the 16th century, the 'Little Ice Age' had set in, and viticulture declined countrywide. In the Industrial Warm Phase, however, viticulture has become well established in the Weald, which may now almost be regarded as the heartland of modern English viticulture.

91

Not surprisingly the first tourist wine trail was started in the Weald (http://village-net.co.uk/customers/englishwine/tours.html)

3.2 The southwest of England

There is little evidence of early viticulture in Cornubia, as the southwest of England was called in pagan times, and is so called to this day by geologists. Though the climate is mild it enjoys a high rainfall. Now, however, the *Côte de Cornubia* is developing along the southern slopes of the mammillated granite spine of Devon and Cornwall from the Exe valley down to the Scilly Isles. Writing half a century ago Hyams (1949) mused *'who will investigate the possibility of turning these wastelands into vineyards?'* The vineyards are planted on a variety of rock types, but are principally on well-drained south-facing slopes of fractured Devonian and Carboniferous slates geologically similar to the vineyards of the Rhine. Examples include Dartmoor, Follymoor, Yearlstone, Beenleigh Manor and Sharpham. The latter was described in some detail in Chapter 5.

3.3 The winelands of the Midlands and northern England

North of the Thames Valley there are few vineyards at present. It is unlikely that the Midland shires will become major centres of viticulture, even if global warming develops. The terrain is largely flat, with occasional southerly-sloping hillsides. The bedrock consists of sedimentary strata of divers types and ages, broadly dipping to the south-east. Over large areas of the Midland plain, however, a range of superficial peri-glacial and glacial deposits, that include 'Boulder Clay', overlies the sedimentary rocks. 'Boulder Clay', as its name suggests, is heavy clay within which are embedded boulders transported by glaciers from as far as Scotland and Scandinavia. This is not a suitable soil for viticulture.

4. Prospective winelands

4.1 The North of England

To the north of the English Midland Plain, however, lies the Pennines, the spine of northern England. It may not be many decades before fine wines are produced from south-facing slopes of the Derbyshire Peak District, where the Carboniferous Limestone produces terrain and soils akin to those of parts of Greece. The Lake District may also have a future, with its well-drained slopes of fractured slates and volcanic tuffs. The prime vineyards will be on the north-western shores of Wast Water and Ullswater, where southeasterly slopes will enjoy the effects of solar reflection from the waters of the lakes below.

4.2 Scottish winelands

Finally, if global warming occurs, it will be the turn of Scotland. Here the *Côtes d'Ecossaise* will flourish, and wine replace whisky as the refreshment of choice. Scotland consists of rocks of many ages and types, but Pre-Cambrian and Lower Palaeozoic metamorphic gneisses and schist predominate, especially in the northern highlands. First vineyards will be planted on the Lower Palaeozoic slates and greywackes of the south-facing slopes of the Southern Uplands. They will be followed by vineyards planted on Dalradian metasediments on south-facing slopes of the Grampians.

The prime Scottish wine estates, however, will be along the north (i.e. south-facing) slopes of the Great Glen where the geology is similar to the Cape vineyards of South Africa. The midge-misted mountains of the metamorphic Moine Series will provide terrain akin to the Bokkefeld meta-sediments of the Stellenbosch vineyards. The intermittent granite intrusions will provide terrain similar to the Paarl region. The prime estates will be on the north shores of the lochs (Loch Ness, Loch Lochy and Loch Linnhe) where the sunny south-east facing slopes will receive enhanced radiation reflected from the waters of the lochs below. There will be tourist wine trails along the Great Glen on land. Luxury cruises, similar to those of the present-day Rhine and Mosel, will sail along the lochs and the linking Caledonian Canal between Inverness to Oban. Refreshed tourists will report more sightings of the Loch Ness monster than ever before (Fig. 7.9).

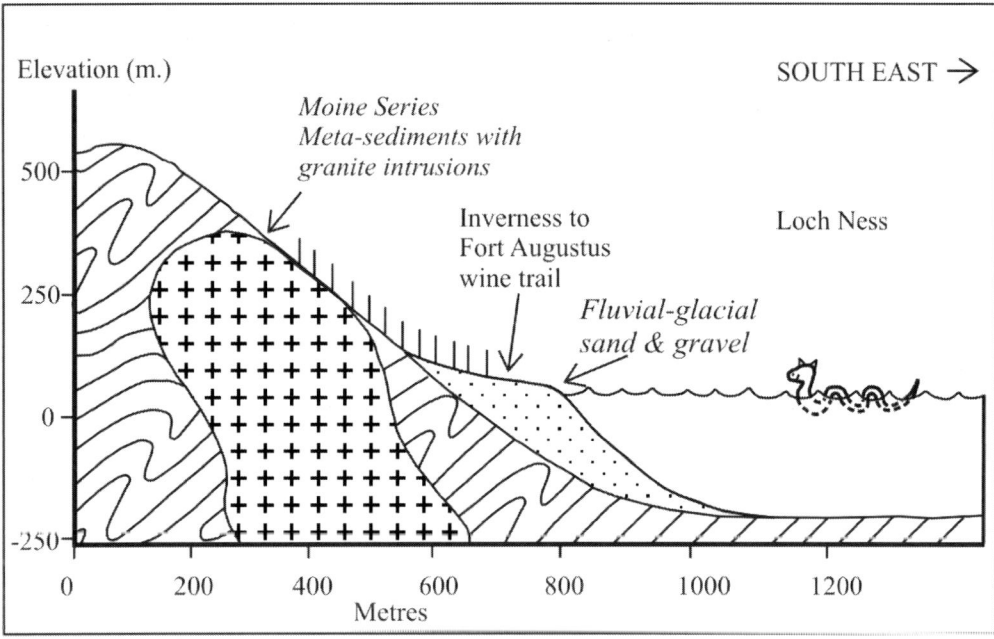

Figure 7. 9 Geological cross-section through the north shore of Loch Ness in the Great Glen, Scotland. With rocks similar to those of the Cape of South Africa, and a sunny south easterly aspect aided by solar reflection from the waters of the lochs, the north side of the Great Glen will host the premier winelands of Scotland, if global warming continues.

Thus geology may be used not only to explain the distribution of present-day vineyards, but also to predict the location of winelands in a future warmer climate (Table 7.3). If present-day climatic predictions are to be believed by the end of this century the Alps will be the rice bowl of Europe, with Swiss peasants tending rice paddies on the slopes of the Matterhorn. Burgundy will be noted for its dates, and Bordeaux for its figs. Britain will be a major wine producer. Even the distinctive South African pinotage grape may grow here. This is the variety of which it is said that *'its juice is extracted from women's tongues and lion's hearts. After a sufficient quantity one can talk forever and fight the devil.'* The vineyards of Britain will be juxtaposed with olive groves. Already olive trees thrive and fructify in the open in southern England. These have been planted by eccentric English gentlefolk, much as their ancestors planted vineyards in the Little Ice Age 300 years ago.

DATE	EVENT	VINE RESPONSE
55 MYA	Deposition of the London Clay	Vines thrive in tropical Britain
2 MYA	Start of the Ice Age	Vines absent from Britain during glacial maxima, present during interglacials
10,000 BP	Start of present interglacial	
2,700 BC	Neolithic culture	First domesticated vines
100 BC	Late Iron Age	Wine imported from Europe
100-300 AD	Roman Warm Phase	Vineyards planted as far north as Lincolnshire
400-950 AD	Dark Age Drop (Saxon Sag)	Viticulture abandoned in Britain
1000-1300	Medieval Warm Phase	Vineyards re-established SE of the Humber-Severn line
1400-1850	'Little Ice Age'	Vineyards restricted to SE England
1850-2000	Industrial Revolution Warm Phase	Vineyards re-established SE of the Humber-Severn line
2000-?	Post-Modern Warm Phase	Vineyards extend N into Scotland?

Table 7.3 Summary chronological table to show the relationship between climate and vines in the British Isles

7.5 THE FUTURE IN FOCUS

As noted earlier the wine vine, *Vitis vinifera,* flourishes today in two belts in the northern and southern hemispheres where the annual temperature averages between $10 - 20^0C$. It is generally recognised, however, that within these isotherms there are cool, intermediate, warm and hot grape varieties (Table 7.4).

During the 20th century British vineyards were largely planted with cool grape varieties suitable for average summer temperatures of between $13-15^0C$. Within the last two decades, however, intermediate varieties in the $15-17^0C$ range have been planted. In southeast England the 'Holy Trinity' of Pinot Noir, Pinot Meunier and Chardonnay now produce medalliferous sparkling white wines. Some heroic vineyard owners have already planted Merlot, albeit in polytunnels.

←COOL →	←INTERMEDIATE→	←WARM →	← HOT →
$13 - 15^0C$	$15 - 17^0C$	$17 - 19^0C$	$19 - 24^0C$
Muller-Thurgau	Pinot Noir	Merlot	Raisins
Reisling	Pinot Meunier	Viognier	Currents
Bacchus	Chardonnay	Syrah	Sultanas
Dornfelder	Sauvignon Blanc	Cabernet Sauvignon	Legbe[x]
Rondo	Cabernet Franc	Grenache	
Regent	Semillon	Zinfandel	

Table 7.4 This shows the grape varieties suitable for cool, intermediate, warm and hot zones. The temperatures are for average summer growing season. [x]Legbe is a wine made from the sap of the date palm, presently produced in the Magreb.

94

In 1900 only a small part of the UK enjoyed average annual temperatures above 10^0C. (Figure 7.10 Top). There were only a few isolated areas in southern England and Wales that were suitable for viticulture. These areas included the Thames Valley, parts of the Weald, the Hampshire Basin and Cornubia. Note that the southern coast of Wales, where the Marquis of Bute planted his vineyard in 1875, was just inside the critical 10^0 C. isotherm. A century later, by 2000AD, most of southern England was above the annual average 10^0 isotherm, except for the Chilterns, Exmoor, Dartmoor and Bodmin Moor. Much of lowland Wales also lay above the critical threshold temperature (Fig. 7.10 Lower).

The Hadley Centre, which is the climate research branch of the UK Meteorological Office, has used data from the Intergovernmental Panel for Climate Change (The IPCC) to predict climate change as far ahead as 2080. The Hadley Centre concludes that summer temperature in south-eastern England will increase by some 5^0C. The increase will only be some 2^0C to the west and northwards to Scotland due to the maritime influence of the Atlantic Ocean. These predictions can be used to fine tune the wild guestimates of the future of British viticulture made in the first edition of this book. Figure 7.11 (Top) shows that by 2080 all of England, except for Dartmoor, the Pennines and Lake District, will be suitable for viticulture, so to will all of Wales, apart from its highland hinterland.

It is possible, however, to make a still more refined prediction of the impact of climate change on British viticulture. Integrating the predicted average summer temperature for 2080 with the temperature ranges for cool, intermediate, warm and hot grape varieties already shown in Table 7.4, it is possible to delineate where different grape varieties may be grown within the UK (Figure 7.11 Bottom).

Whether these predictions come to pass will depend on many factors, climatic and political. The predictions of future climate become ever more robust as the years roll by, more data accumulate, and trends in temperature and precipitation more focussed. There seems little chance now that politicians have the will to make the major changes required to limit the production of greenhouse gas emissions, though, in the western world at least, the masses largely accept the need for restriction. World population continues to rise, while at the same time food, water and energy become scarcer. Large population migrations accompanied by a break down in law and order will follow. It may well be that by 2080 the inhabitants of what is now the United Kingdom may be feral gangs fighting over abandoned supermarkets foraging for edible detritus long past its sell by date. The life style of the hunter gathers of the Mesolithic age will return.

In the short term, however, the future is bright for the UK wine industry. When the first edition of this book was written in 2003 the most northern vineyard in England was Mount Pleasant, in Lancashire, at Latitude 54^0 5'. Now, in 2008, the northern limit of viticulture has advanced to Acomb, Yorkshire, at Latitude 55^0 0'. This vineyard is only 3 miles south of Hadrian's Wall. So there are advantages to a warmer Britain. We should plant vineyards and enjoy our liquid assets while we may.

Already there is a tea plantation at Tregothnan in Cornwall, an almond grove near Honiton, Devon, and olive trees are ubiquitous in gardens across southern England (Fig. 7.12). Olives can now be easily cultivated if they treated like teenagers. They should be ignored as far as possible, not over fed, and never ever given too much to drink.

Though climate will always control the regional distribution of viticulture, the location of individual vineyards will be controlled by the way in which geology controls topography and soil.

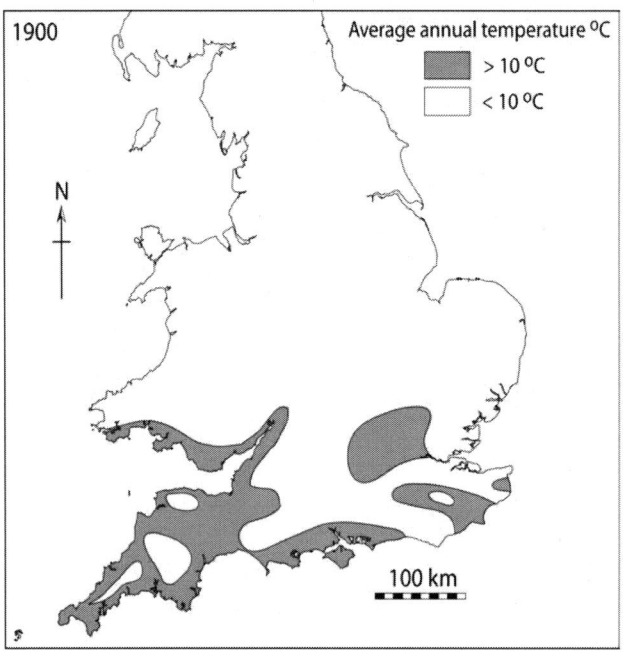

Figure 7.10 Above and below, maps to showing the increase between 1900 and 2000AD in the extent of the United Kingdom where the annual average temperature required for successful viticulture exceeds 10^0C

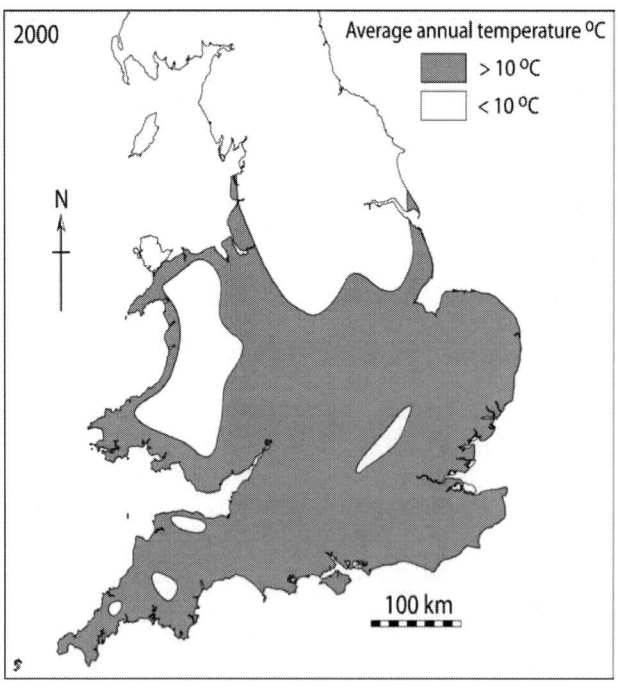

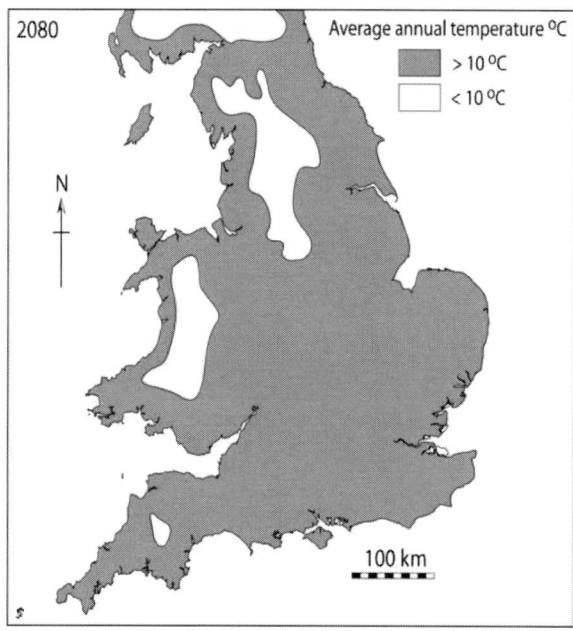

Figure 7.11. Above: Predicted average annual average temperature for 2080 calculated by the Hadley Centre using IGCP global temperature predictions. Below: Suitability for growing various grape cultivars in 2080 according to predicted average summer temperature.

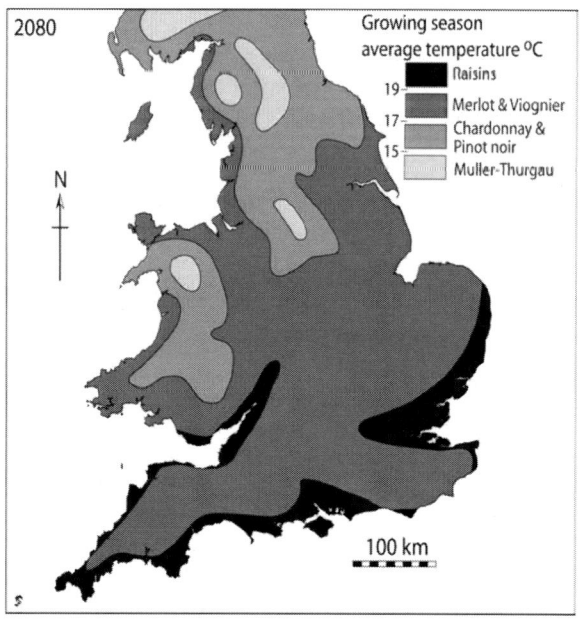

Figure 7.12. Surrey peasant enjoying a glass of local wine accompanied by a dish of olives from his adjacent olive grovelet.

7.5 BIBLIOGRAPHY

Books on climates and palaeo-climate that are accessible to the general reader are thin on the ground, but try:

Burrows, W.J. 2007. Climate Change: A Multisciplinary Approach. Cambridge University Press. Cambridge. 390pp.

Cowie, J. 2007. Climate Change: Biological and Human Aspects. Cambridge University Press. Cambridge. 487pp.

Houghton, J.H 2004. Global Warming. The Complete Briefing. (3rd Edn.) Cambridge University Press. 300pp.

Hubert, B.T., Macleod, K.G., and Wing, S.T. (Editors) 2000. Warm Climates in Earth History. Cambridge University Press. Cambridge. 462pp.

Solomon, S. et al. 2007. Climate Change 2007. The Physical Basis. Report of the Intergovernmental Panel on Climate Change. Part. 1. Cambridge University Press. Cambridge. 1009pp.

7.6 REFERENCES

Columella, 67. De Re Rustica.

Cuffey, K.M. and Brook, E.J. 2000. Ice Sheets and the Ice-Core Record of Climate Change. In: Earth System Science. Jacobsen, M. C., Charlson, R. J., Rodhe, H. and Orians, G. H. (eds.) Academic Press. London. pp. 459-497.

Denning, Lord. 1992. Foreward. In: The Magistrate as Chairman. Butterworths. London. xi-xiv.

Field, D. and Cotton, J. 1987. Neolithic Surrey: a survey of the evidence. In: The Archaeology of Surrey to 1540. (Bird, J. and Bird, D.G. eds.) Published by the Surrey Arch Soc. Guildford. pp. 71 – 96

French, W. 2002. Global changes in the Earth and the Galactic year. Geologists' Association Mag. 3. 1. 17

Gibbard, P.L. 1985. The Pleistocene History of the Middle Thames. Cambridge University Press. Cambridge. 155pp.

Hannebuth, T. et al. 2000. Rapid flooding of the Sunda Shelf: a late glacial sea level record. Science. 288. 1033 – 1035

Hardy, J. 2003. Climate Change, causes, effects and solutions. J. Wiley. Chichester. 256pp.

Houghton, J.T. and Jenkins, G.L. (eds.) 1990. Climate Change. The International Panel on Climate Change Assessment. Cambridge University Press. Cambridge. 403pp.

Hyams, E. 1949. The Grape Vine in England. John Lane. London. 209pp.

Jacobson, M.C., Charlson, R. J., Henning, H. and G. H. Orians. 2000. Earth System Science. Academic Press. London. 527pp.

Jones, G.V. and Davis, R.E. 2000. Using a Synoptic Climatological Approach to Understand Climate/Viticulture Relationships. International Jl. of Climatology. 20. 813-837.

Lamb, H.H. 1995. Climate History and the Modern World. Routledge. London. 433pp.

Le Roy Ladurie, E. 1983. Histoire du Climat Depuis l'An Mil. Flammarian. Paris. 285pp.

Lovelock, J E. and Margolis, M., 1974. Atmospheric homeostasis by and for the biosphere. Tellus. 26. 1 – 10.

Manley, G. 1952. Climate and the British Scene. William Collins Sons & Co. Ltd. Glasgow. 382pp.

Merritts, D., De Wet, A. & K. Menking. 1998. Environmental geology: an E arth system science approach. W H Freeman & Company. New York. 452pp.

Nansen Arctic Drilling Program Science Committee (1992) The Arctic Ocean Record: Key to Global Change. Polarforschung 61/1 102pp.

Payne, S. 2002. Word of the Month. Global Warming. Wine Society News. Stevenage. p.4

Siegert, M. 2001. Ice Sheets and Late Quaternary Environmental Change. J. Wiley. Chichester. 288pp.

Skinner, B.J. & Porter, S. C. 1995. The Dynamic Earth. (3rd edn.) J Wiley & Sons. New York. 563pp.

Steiner, J. 1967. The sequence of geological events and the dynamics of the Milky Way Galaxy. Jl. geol. Soc. Australia. 14. 99-132.

Tate, A.B. 2001. Global Warming's Impact on Wine. Jl. Wine Research. 12. 95-109

Veevers, J.J. 1990. Tectonic-climatic supercycle in the billion-year plate-tectonic eon: Permian Pangean icehouse alternates with Cretaceous dispersed-continent greenhouse. Sed. Geol. 68. 1-16.

Walker, G. 2002. Snowball Earth. Bloomsbury. London. 269pp.

Watson, R.T, and the Core Writing Team (eds.) 2001. Climate Change 2001: Synthesis Report. Cambridge University Press. Cambridge. 397pp.

Winchester, S. 2001. The Map that Changed the World. Penguin. London. 338pp.

GEOLOGICAL AND OENOLOGICAL GLOSSARY

Alluvium, superficial deposit (q. v.) laid down by rivers.

Ampelography, the science of grape-bearing vines.

Anticline, an upfold of sedimentary rocks (q. v.), e.g. the Weald of southeast England.

Arpent, a Norman measure of land, slightly under one acre, approximately 0.3 of a hectare, widely used in the Domesday Book as a unit for measuring vineyards

Auxerrois, a vine variety from Alsace that does well in England

Bacchus, both a variety of grape of German origin, that thrives in England, and the name of the Roman god of wine

Botrytis cinerarea, the 'noble rot', a fungus that attacks grapes in favourable autumnal weather conditions, enhancing the sugar content and producing dessert wines of high quality.

Boulder Clay, a glacial deposit consisting of clay containing sub-angular stones of many sizes scattered randomly throughout, producing heavy soils throughout East Anglia and parts of the English Midlands

British wine, a term applied to wine made in the UK from grape juice imported from who knows where. But note, with wines produced in the Channel Islands, England and Wales, and shortly in Scotland, the term may become used in a dual sense

Chalk, a white friable limestone, highly porous, but of low permeability, unless fractured. Important for viticulture in France and England.

Champagne, an area of rolling chalk downs and escarpments in north-eastern France that produces the eponymous wine by the *méthode Champenoise*. The latter may be exported, the former, applied to a wine, may not.

Chardonnay, popular white wine variety in the old world and the new that struggles in England.

Chert, silica (SiO_2) that forms in sedimentary rocks (q. v.) after deposition. It occurs replacing fossils and in nodules and layers. Flint (q. v.) is a variety of chert found in Chalk (q. v.)

Chlorosis, disease of the vine in which the leaves turn yellow. Caused by iron deficiency in limestone soils.

Clay, a very fine-grained sediment, generally porous, but impermeable.

Clay-with-flints, a superficial deposit (q. v.) of uncertain origin, composed of clay with scattered flints (q. v.) which blankets the top of the Downs (q. v.) across much of southern England

Crater, a Roman wine cup

Creation science, an oxymoronic theory that the earth was created in 6 days, in defiance of nearly 2 centuries of observations that rock sequences contain fossils of ever increasing biodiversity and complexity from bottom (oldest) to top (youngest).

Cultivar, a posh word for a vine variety, see also Varietal

Dornfelder, a red wine vine variety from Germany that does well in England

Downs, open rolling countryside with dry valleys (q. v.) and scarps composed of Chalk limestone, comparable to the Champagne area of France (q. v.)

Drift, an obsolete term applied to superficial deposits (q. v.), originally thought to be detritus left by the Noarchian flood.

Dry valley, a valley that lacks a river, a characteristic feature of the Chalk downlands (q. v.). Dry valleys formed in periglacial episodes when summer melt water eroded river valleys, being unable to percolate through frozen ground. Dry valleys provide sheltered locations for modern vineyards

Dionysus, the ancient Greek god of wine, equivalent of the Roman Bacchus (q.v.)

Elbling, red and white wine vine varieties possibly descended from the pre-Roman *Vitis alba*, grown in England

English wine, wine made from grapes grown in England. Distinct from British wine (q. v.)

Evolution, not a theory, but a synthesis of two centuries of geological observations that fossils reveal increasing biodiversity and complexity from the lowest (oldest) to the highest (youngest) sedimentary strata. Evolution has been explained by several theories including Lamarkism, Darwin's theory of evolution by natural selection and poly-Noarchism – that successive assemblages of fossils of increasing complexity indicate successive creations and floods.

Falx vinitoria, a small pruning-hook used in Roman and medieval times for viticulture.

Fault, a break in rock with lateral movement.

Flint, a variety of chert (q. v.) that forms nodules, often in layers, in the Chalk (q. v.), and in Clay-with flints (q. v.). Flint has no connection with the 'flinty' taste in wine detected by hypersensitive palates

Fracture, a break in rock without any lateral movement. Fractures increase the permeability (q. v.) of rock.

Geology, the study of the Earth, as opposed to Theology, the study of Heavenly things, as defined by Richard, Bishop of Durham in the 14th century.

Gneiss, (pronounced 'nice') a metamorphic rock (q. v.) of granitic composition, the end result of subjecting diverse rocks to high temperatures and pressure.

Granite, an igneous rock composed of the minerals quartz, feldspar and mica. It forms from the slow cooling of magma (q. v.).

Greensand, sandstone containing the green mineral glauconite that weathers out at the surface to rust (iron oxide). Greensands are common in France and England and host vineyards in the Loire, and hosted vineyards in the abandoned winelands of the Surrey Hills in the Little Ice Age.

Huxelrebe, a cross-breed vine variety of German origin widely used for making white wine in England

Igneous rock, one formed from the cooling of magma (q. v.) either cooling slowly underground, giving rise to coarsely crystalline rocks like granite, or erupting at the Earth's surface in volcanoes as lava.

Law of superposition, a fundamental principle of geology, that in a sequence of sedimentary strata (q. v.) the higher strata are younger than the lower strata.

Limestone, a sedimentary rock (q. v.) composed of lime (calcium carbonate) of which chalk (q. v.) is a particular variety

Lithification, the learned term given to describe the process whereby soft sediment is turned into solid rock by temperature and pressure during burial.

Madeleine Angevine, a white wine vine originating in Germany with a complicated parentage and etymology that does well in England. It has a synonymous table grape doppelganger

Magma, molten material from beneath the earth's surface that cools to form igneous rocks (q. v.).

Marl, a calcareous (lime rich) clay.

Metamorphic, a group of rocks formed by high temperature and/or high pressure, from pre-existing rocks, such as slate, schist and gneiss.

Merlot, a red wine vine of French origin, as yet grown by only the most heroic of English vinearons.

Muid, a Norman liquid measure used for wine, equivalent to about 36 gallons or 164 litres.

Muller-Thurgau, a sort of Swiss-German vine variety cross of complicated parentage, widely grown in England for white wine.

Oenology, the study of wine

Ortega, a German cross-breed wine variety widely used in England for white wine

Pastinatio, type of vine planting described in Roman texts, and found by archaeologists in English Roman vineyards.

Permeability, the property of a rock whereby fluid can flow through it. Indicative of interconnected porosity (q. v.)

pH, a measure of acidity/alkalinity of which 0 is the most acid, 7 is neutral and 14 the most alkaline

Pinot gris, a white wine vine variety that makes Botrytised (q. v.) dessert wines

Pinot Meunier, a white wine vine variety used in Champagne (q.v.) that struggles in England

Pinot noir, a red wine vine variety, widely planted, but temperamental in the English climate

Porosity, the storage capacity of a rock, generally expressed as a percentage. Pores occur in many ways, some primary, e.g. the spaces between sand grains, others secondary, e.g. fractures (q.v.). At shallow depths pores may be full of air. Below the water table (q.v.), pores are full of water. Interconnected pores make a rock permeable (q. v.), and allow the flow of water. Pores in very fine sedimentary rocks, such as clay (q. v.) and chalk (q. v.) may contain water, but be impermeable due to surface tension around the grains. Fractures, however, may impart permeability to such rocks.

Quartzite, a metamorphosed sandstone (q.v.), composed almost entirely of the mineral quartz (silica) commonly low in nutrients and highly fractured, and hence permeable

Reichensteiner, a white wine vine German cross-breed widely planted in England

Reisling, a ubiquitous German white wine vine variety.

River terrace, a flat area on the flank of a river valley above the level of the present flood plain. River terraces are relicts of past flood plains eroded by the river cutting down to its present level. Being flat and well-drained, river terraces are often good for viticulture.

Rondo, a red wine grape that has migrated, with the setting sun, along the northern climatic limit of viticulture from its origin in Manchuria to England.

Sand, sediment formed of sand-sized particles, generally of the mineral quartz, silica.

Sandstone, lithified (q.v.) sand.

Schist, metamorphosed clay (q. v.) transitional between slate and gneiss (q. v.).

Schönburger, a white wine vine German cross that does well in England

Sedimentary rock, rock formed by the lithifaction (q. v.) of deposits of grains of clay, sand and gravel. Also includes rocks, such as salt, gypsum and dolomite that form by evaporation or replacement of pre-existing sediments.

Seyval blanc, a white wine vine French cross-breed widely planted in England

Siegerrebe, a white wine vine derived from Madeleine Angevine (q. v.) that does well in England

Shale, lithified (q. v.) clay (q. v.)

Slate, a low grade metamorphic rock formed when clay is subjected to high pressure and temperature, en route from shale (q. v.) to schist (q. v.)

Soil, a mixture of organic matter, alive and dead, and weathered rock detritus.

Stratum, single, strata, plural, layer(s) of sedimentary (q. v.) rock.

Superficial deposit, shallow sedimentary detritus unrelated to the bedrock that it overlies, includes alluvium, deposited by rivers, and a diverse suite of glacial and periglacial deposits. May be more important to soil (q. v.) and viticulture than the bed rock beneath.

Syncline, a downfold of sedimentary rocks (q. v.)

Terroir, a quasi-mystical quintessesentially French term beyond the comprehension of AngloSaxons, used to describe the synergy of geology and climate that effect wine character.

Terroiriste, a true believer in terroir (q.v.)

Unconformity, a discontinuity between two rock masses due to non-deposition or erosion. The upper is sedimentary, the lower of any origin. Where unconformities separate sedimentary rocks the lower sequence often dips at a steeper angle than the upper strata. An unconformity generally indicates a break in deposition, and hence a gap in time.

Varietal, a distinctive term for wine made from a single grape variety.

Vinearon, a wine maker (English), a term coined by Edward Hyams (1949)

Vigneron, a wine maker (French)

Vine variety, a subspecies of *Vitis vinifera* (q. v.), e.g. Chardonnay, Pinot Noir, etc..

Vinification, the metamorphism of grape juice into wine. As opposed to viticulture (q. v.) the cultivation of vines.

Viticulture, the cultivation of vines, as opposed to vinification (q. v.), the metamorphism of grapes into wine

Vitis vinifera, the scientific name for the common vine.

Water table, surface above which the pores in rock are dry and air-filled, and below which the pores are full of water

Weathering, the breakdown of rock by physical, chemical and biological processes. The prelude to soil formation

Wineland, a wine-growing region defined by its geology, and often period of cultivation.

Xenoenophobia, fear of foreign wine

Xenoenophilia, enjoyment of foreign wine

INDEX

Vineyards are printed in **Bold Face**

SOME USEFUL WEBSITES

Most vineyards have their own websites that can easily be found on a search engine, but here are some useful links:

Denbies Winery: www.denbiesvineyard.co.uk

English Wine Centre: www.englishwinecompany.co.uk

English Wine Company: www.englishwineproducers.com

English Wine Producers: www.englishwineproducers.co.uk

English-wine.com: www.english-wine.com

Englishwine.com: www.englishwine.com

Petravin: www.petravin.co.uk

Richardson's English Wine Site: www.sol.brunel.ac.uk/~richards/wine

UK Wine Site: www.easyweb.easynet.co.uk/~andie

United Kingdom Vineyard Growers Association www.ukva.org

Vinopolis: www.vinopolis.co.uk

Viticulture UK: www.viticulture.co.uk

Wine on the Web: www.wineontheweb.com